REBUILDING
THE PULP AND PAPER
WORKERS' UNION,
1933–1941

ℑ Twentieth-Century America Series

REBUILDING
THE PULP AND PAPER
WORKERS' UNION,
1933–1941

Robert H. Zieger

The University of Tennessee Press / Knoxville

ℒ Twentieth-Century America Series
DEWEY W. GRANTHAM, General Editor

The paper in this book meets the guidelines for permanence
and durability of the Committee on Production Guidelines for
Book Longevity of the Council on Library Resources. Binding
materials have been chosen for durability.

Library of Congress Cataloging in Publication Data
Zieger, Robert H.
    Rebuilding the pulp and paper workers' union, 1933–
1941.
    (Twentieth-century America series)
    Bibliography: p.
    Includes index.
    1. International Brotherhood of Pulp, Sulphite, and
Paper Mill Workers—History. I. Title. II. Series.
HD6515.P33Z53 1984   331.88'176'0973   83-10227
ISBN 0-87049-407-4

Publication of this book has been aided by a grant from the
National Endowment for the Humanities.

For Robert,

who grew up with the Pulp Workers

Contents

Preface   ix

1   Introduction   3

2   The Pulp and Paper Industry   15

3   John Burke and the Union   45

4   Using the National Industrial Recovery Act
in the Paper Industry   66

5   Organizing in the Thirties   95

6   Moving Ahead on All Fronts   127

7   A Union among Unions   168

8   Change and Continuity   194

9   Legacy   214

Bibliographical Essay   223

Index   233

Illustrations

Logs being prepared for pulping   *page* 30

Paper workers toiling at the "dry" end of a fourdrinier   31

Emblem of the International Brotherhood of Pulp,
Sulphite, and Paper Mill Workers   32

John P. Burke   33

1939 convention in St. Paul   102

International headquarters in Fort Edward, New York   103

Preface

Those in the category labelled "Other" [in a survey of pre-World-War-I German workers] included 46 percent of Ruhr workers. They are the forgotten men and women. . . . Except for their appearance in the census it would be difficult, if not impossible, to take them properly into account. . . . their fortunes and misfortunes have not found their way into tales of workers' struggle. . . . To bear one's fate silently is no way to get into history books, even obscure and dull ones. With some regret—and unsatisfied curiosity—it will be necessary to leave them aside.

—Barrington Moore, Jr., *Injustice*

It cannot fairly be said that pulp, paper, and converted paper workers have been entirely ignored in the writings of scholars. Still, their story, and the story of their union, remains largely unknown, even by those whose primary interest is American labor history. This statement is especially true—and especially lamentable—in connection with the 1930s, the period of the paper industry's leap toward unionization. Extant works, largely by labor economists, either deal with the period before the depression or touch upon the emergence of the International Brotherhood of Pulp, Sulphite, and Paper Mill Workers in the 1930s tangentially, as background for more immediate concerns. On the other hand, the historians who have done so much to bring the turbulent decade alive have ignored an industry that contained more than 400,000 workers.

The people who toiled in occupations falling under the jurisdiction of the Brotherhood performed a variety of tasks, most of them carrying the designation semiskilled or unskilled. The union included in its ranks woodsmen who cut and hauled pulpwood in remote forests; workers who manhandled the logs, stripped them of their bark, and ground and chipped them into pulpable form; and still other workers who made acid and tended the great cookers and beaters in the pulp mills. In paper mills, Brotherhood members cleaned the screens and felts that the giant fourdrinier machines used to extract the water from the pulp to produce paper. They swept the floors, collected the "broke," or refuse, and moved the rolls of paper after they wound off the machines, readying them for finishing operations and shipping.

By the 1930s, the vast converted paper industry, employing more than 200,000 workers, began to open to unionization. Workers in bag and tissue mills and in container plants, scoring and stitching machine operators in corrugated box shops, makers of construction materials, men and women who fashioned drinking straws, muffin papers, plates, cups, cereal boxes—all of these, and many whose occupations fell between the interstices of statistical categories, were eligible for organization by the Brotherhood. And in the charged atmosphere of the thirties, thousands of them joined the union.

A combination of circumstances underlay the Brotherhood's quiet success during the period. The union's ability to survive through the grim twenties paid off in the heady enthusiasm of the early New Deal. The dramatic changes in federal labor policy, as evidenced particularly by Section 7(a) of the National Industrial Recovery Act and the National Labor Relations Act, provided decisive impetus. And the vigorous challenge of the Congress of Industrial Organizations triggered gains for the American Federation of Labor unions in the paper industry. I hope that the assessment of these forces in the chapters that follow will encourage students of the 1930s and of the American labor movement to take the Brotherhood and its workers more fully into account.

Research for this work began in the shadows of Wisconsin's paper mills and continued through an interlude in northeastern Kansas. I wrote the book largely in Detroit, a fitting place for the study of any industrial union, even one that flew the AFL banner. During the research and writing, I incurred many debts. As always, Horace Samuel Merrill was a source of encouragement and invaluable advice. Le Roy Ashby and Patrick Maney read most of the manuscript and provided both the needed criticism and the even more needed support. Charles Hyde and Burton Kaufman offered valuable suggestions. Marcia Brubeck of the Guilford Group strengthened the manuscript through her excellent copy editing. Ginny Corbin typed it with skill and dispatch. The National Endowment for the Humanities, the American Philosophical Society, Kansas State University, and Wayne State University supported the research. Of course, the manuscript's limitations remain my sole responsibility.

Several excellent libraries and other depositories provided me with materials and assistance. The staff of the State Historical Society of Wisconsin was unfailingly helpful throughout this project. The staffs of the National Archives and Records Service, the Catherwood Library of Cornell University, the Archives of Labor History and Urban Affairs of Wayne State University, the AFL-CIO Library in Washington, and the U.S. Department of Labor Library all supplied essential research material and first-class assistance. I owe a special debt to Richard Strassberg of the Labor-Management Documentation Center, Cornell University, for bringing to my attention the Westvaco Papers and for other services. The United Paperworkers' International Union generously supplied me with microfilmed and photoduplicated materials and other assistance. My friend Barbara Hudy Bartkowiak rendered important research assistance in locating Wisconsin materials.

George W. Brooks of Cornell University provided me with important information and insights, especially during a telephone interview in 1979. I also interviewed Nicholas Vrataric and benefited from his vast knowledge of unionism in the paper industry. Thanks, too, to Clarence Farmer of Local 675, Covington, Virginia, for permitting me to use, in connection with chapter 9, material gathered during an interview in 1980 for another project. Thurmon Reynolds, also of Covington, provided me with a photograph of the Pulp, Sulphite, and Paper Mill Workers' convention of 1939 and helped to identify the individuals depicted. Louis Gordon, Monty Byers, and Faye Hale of the United Paperworkers' International Union, and George W. Brooks also helped locate and identify photographic materials. I owe special thanks to Judson MacLaury of the U.S. Department of Labor for help with photographs, and to Ronald Fahl of the Forest History Society, who provided me with good copies of most of the photographs in this book. As usual, the advice and support of my wife, Gay Pitman Zieger, was, literally, indispensable.

Oak Park, Michigan   RHZ
October 1, 1982

REBUILDING
THE PULP AND PAPER
WORKERS' UNION,
1933–1941

1

Introduction

It is the hardest job in the world to organize working
people. I know, because I have been at it for more than
twenty-five years.

> —John P. Burke, president-secretary, Inter-
> national Brotherhood of Pulp, Sulphite,
> and Paper Mill Workers, 1917–65

ETWEEN 1917 AND 1965, JOHN P. BURKE ADDRESSED THE DELE-
gates at every convention of the International Brotherhood
of Pulp, Sulphite, and Paper Mill Workers in his capacity
as president-secretary. At the beginning of his speech, he almost
invariably reviewed the union's troubled but ultimately edifying
history. He did so not only because new delegates too often knew
little of their own organization's origins and struggles but also because he believed that other chroniclers of the labor movement,
hypnotized by the largest and most powerful unions and capti-
vated by the rhetoric of the Lewises and Reuthers, overlooked such
organizations as his beloved Brotherhood. "The trouble with labor
history in this country is this," he told delegates to the eighteenth
convention in St. Paul in March 1939: "Intellectuals have written
labor history, and they have ignored completely the struggles of
the smaller organizations."[1] In consequence, every gathering of
the union found John Burke setting the record straight.

The revival of Burke's union in the 1930s does indeed reveal aspects of the history of the American labor movement that are
often slighted when we concentrate on the more dramatic aspects
of that critical decade. The Brotherhood's experiences with work-

---

1. Brotherhood, *Report of Proceedings . . . 18th Convention* (1939), 11.

ers' militancy, with government involvement in unionization, and with the ongoing traditions of the labor movement all illuminate some of the problems that the next generation of scholars and critics would come to see as products of the "maturation" of the labor movement that followed World War II. For example, union organizers often found the vaunted militancy of depression-era workers difficult to comprehend and to channel because it was sporadic. Directing that volatile spirit into stable organizations required the organizers to grapple with time-honored practices and institutions of the ongoing labor movement. Many workers viewed unionism in the 1930s as consisting simply of the per capita tax, the monthly dues demanded by the union in return for the services it rendered. Organizers and officers spent countless hours devising schemes to convert a group of workers with a spark of interest in the union into a disciplined body of regular dues payers. At the same time, Brotherhood leaders had to concern themselves with jurisdictional questions associated with the Brotherhood's membership in the American Federation of Labor (AFL), so that they functioned in part as laborite entrepreneurs, grasping opportunities and fighting off the challenges of rival unions.

Pulp and paper workers often viewed the union diffidently and even suspiciously. Many originally joined primarily to express support for Roosevelt's National Industrial Recovery Act. Others explored both the AFL and the Congress of International Organizations (CIO), seeking the best deal. Many found the Brotherhood's cautious adherence to traditional AFL practices uninspiring. Some considered the burgeoning legalistic structure of industrial relations, which the international union both decried and embraced, frustratingly bureaucratic. For the veteran leaders of the pulp and paper workers, and for some of the younger organizers who were earning their spurs, the union represented a lifetime commitment, an end in itself, while for large numbers of new recruits it was, more simply, a means to a better life.

The individuals involved with the International Brotherhood of Pulp, Sulphite, and Paper Mill Workers in the 1930s experienced many confrontations. Whether they were long-tenured officers, newly appointed organizers, or merely members, they battled employers, faced often dilatory government agents, and confronted rival labor organizations of both the AFL and the CIO. More constant and more fundamental, however, was still another confrontation, that of worker versus union.

John P. Burke believed that so-called experts invariably failed to comprehend the difficulty of building and sustaining labor organizations in the United States. "It is so hard to organize workers," he once remarked. Armchair socialists "think that the reason we have not ten or twelve million workers organized is because the organizers are all lazy and do not do anything."[2] The true reason, in his view, lay in the tendencies of workers to place private concerns, religious sentiments, political commitments, even gambling, hunting and fishing, and socializing, before unionism.[3]

Indeed, organizers have rarely found their work easy in the United States. Unions have grown grudgingly and sporadically —when they have grown at all. Their proponents have, with a few exceptions, found themselves obliged to appeal to recruits on the narrowest grounds of practicality and to avoid explicit ideologies. Even today, barely 20 percent of the nonagricultural labor force is organized. At their highest point of relative membership in the mid-1940s, the unions, their constituencies boosted by maintenance-of-membership devices and by the war's favorable effects on employment, could count one-third of American workers as members. In 1932, by contrast, enrollment had slipped beneath 3,000,000 — barely 8 percent of the depression-ravaged labor force—and was concentrated in a handful of trades, several of them quite peripheral to the economic mainstream. The intervening years, the decade of the thirties, constituted the most dynamic period of growth in the history of the American labor movement. Workers had surged into labor organizations during the 1880s, of course, and again in the months of World War I, and World War II increased union membership by more than 40

2. John P. Burke to H.W. Sullivan, April 7, 1930, Brotherhood Papers, microfilm, 1930, Reel 1. Complete runs of the union's papers are available at the State Historical Society of Wisconsin and at the Labor-Management Documentation Center, Martin P. Catherwood Library, New York State School of Industrial and Labor Relations, Cornell University. Hereafter I shall cite these records by the date and the reel number of the document. Reel numbers begin with number one for each year.

3. Burke's correspondence throughout the 1930s is filled with these themes. Herbert W. Sullivan to Burke, April 7, 1930; Burke to George Brooks, May 28, 1930; Burke to R.C. Henke, June 1, 1930; Burke to Brooks, Oct. 31, 1930, all on Reel 1. See Voelker, "The History of the International Brotherhood," 93–101.

percent. Still, the thirties were decisive, not only for the labor movement in general, but also for its component parts, the national and international unions. Some of America's great organizations, notably the United Automobile Workers (UAW) and the United Steelworkers of America, were born. Others, most spectacularly the United Mine Workers, were reborn. Some died in the decade—the Amalgamated Association of Iron, Steel, and Tin Workers provides a little-lamented example.[4]

The influence of the thirties also expressed itself in less dramatic ways. Unions such as the United Brotherhood of Carpenters and Joiners and the International Brotherhood of Teamsters grew enormously during the New Deal era, creating new structures, absorbing new kinds of members, and sending forth new leaders, all the while retaining powerful elements of their traditional character. The federal labor unions, the AFL's forgotten stepchildren, surged into prominence for a season. Other labor organizations, such as the one that eventually became the Communications Workers of America, groped their way from company unionism to trade unionism.[5]

In explaining the growth of the labor movement in this period, most observers have identified four central factors, usually in combination: the favorable role played by the federal government; the anger and militancy of ordinary workers; the work of a cadre of highly motivated, ambitious, and effective laborites, many of them radicals; the presence of an ongoing, if weakened and disoriented, mainstream labor movement.

Commentary on the labor history of the 1930s has been spirited, as befits a subject so filled with drama and spectacle.[6] Mem-

---

4. The best studies of the rise and fall of the various unions in the 1930s are Galenson, *CIO Challenge*, and Bernstein, *Turbulent Years*. Neither refers to the pulp and paper industry or to the Brotherhood.

5. On the Teamsters, see Bernstein, *Turbulent Years*, 240–43 and 780–81, and Galenson, *CIO Challenge*, 459–94. On the Carpenters, see Christie, *Empire in Wood*, 287–300. On the federal labor unions, see Zieger, *Madison's Battery Workers*, 6, 18, 33–34, and 44. On the Communications Workers, see Schacht, "Toward Industrial Unionism," and Brooks, *Communications Workers*, 1–64.

6. For an indication of the scope of historiographical debate, see, e.g., Bernstein, *Turbulent Years*, 768–95; David Brody, "Labor and the Great Depression: The Interpretive Prospects," *Labor History* 13 (Spring 1972), 231–44, and "Radical Labor History and Rank-and-File Militancy," *Labor History* 16 (Winter 1975), 117–26; and Green, "Working Class Militancy in the Depression."

oirs and autobiographies[7] have centered on heroic struggles. Films that draw upon the copious newsreel footage produced during the 1930s are vivid reminders of the decade's turbulent history. Thus far, the most exploited of the recently established archival collections have been those relating to the great CIO unions, with their violent origins and tumultuous histories.[8]

Despite this frequent focus on violence, spectacle, and innovation, the system of labor relations that emerged from the decade was a complex and often ambiguous phenomenon. This was a period of trial and error, of passion and caution, of change and continuity. Although hindsight permits some of today's scholars to view the rise of the new unions and supporting legislation as means of adjusting workers to a smooth-running corporate order,[9] conflict between labor and management was bitter, even fanatical, at the time. If millions of workers boarded the union train, millions of their peers turned their backs, clinging to individualism — to their private goals and to hopes for social mobility. The angry militancy that underlay much of the unionization of the 1930s was as sporadic as it was spectacular, the product, some students conclude, as much of carelessness and tactical mistakes on the part of employers as of reflexive working-class solidarity. While some witnesses to the golden days of the thirties dreamed of social transformation and of political and economic reconstruction led by a powerful and radicalized labor movement, others, no less dedicated to the cause of unionism, confined their vision to the procurement of signed contracts with rigorous overtime and se-

---

7. E.g., Alice Lynd and Staughton Lynd, eds., *Rank and File: Personal Histories by Working-Class Organizers* (Boston: Beacon, 1973), Farrell Dobbs, *Teamster Rebellion* (New York: Monad, 1972); Dobbs, *Teamster Power* (New York: Monad, 1973); Henry Kraus, *The Many and the Few: A Chronicle of the Dynamic Auto Workers* (Los Angeles: Plantin, 1947); and titles cited in the bibliographical essay.

8. But see Tomlins, "AFL Unions in the 1930s"; Morris, *Conflict within the AFL*, 136–290; David Brody, "The Expansion of the American Labor Movement: Institutional Sources of Stimulus and Restraint," in Stephen E. Ambrose, ed., *Institutions in Modern America* (Baltimore: Johns Hopkins Univ. Press, 1967), 11–36; and my bibliographical essay.

9. E.g., Staughton Lynd, "The Possibility of Radicalism in the Early 1930's: The Case of Steel," *Radical America* 6 (Nov.-Dec. 1972), 37–64; Lynd, "Guerrilla History in Gary," *Liberation* 14 (Oct. 1969), 17–21; Piven and Cloward, *Poor People's Movements*, 96–180; and Stephen Scheinberg, "The Development of Corporation Labor Policy, 1900–1940" (Ph.D. diss., Univ. of Wisconsin, 1966), 163–87.

niority provisions, the establishment of effective grievance machinery, and the fattening of pay packets.

Nor can the story of the thirties be told simply in terms of conflict between social or radical unionism on the one hand and enduring traditions of business unionism on the other. Whereas the rubric "industrial unionism" once implied radical notions of labor's purposes and functions, by the end of the thirties it was becoming a politically neutral term for the recruiting and representing of workers. Moreover, organizations that adhered closely to the full panoply of "business union" practices were by no means universally bereft of idealism, nor were they led by people unconcerned with the broader aspects of American society. Unions then as now (and perhaps more so then, before jurisdictions and structures had become fixed) were deeply human institutions. In responding as organisms to the stimuli of the highly charged New Deal period, they disregarded the classifications and categories of chroniclers and taxonomists. Much of the scholarly attention that has been directed to the thirties has focused on the behavior of masses of workers or on the activities of a few bold and aggressive leaders. We have neglected the local unions, and we have studied (and only sketchily) few but the very largest and most powerful of the international unions.

The rise of the labor movement in the 1930s involved millions of ordinary men and women. Literally countless local union meetings, picket lines, organizing drives, and government hearings and mediating sessions constituted its quotidian activities. While it was indeed the era of Harry Bridges and John L. Lewis, of Walter Reuther and William Green, thousands of less-known but eager and energetic unionists were also attempting to seize the day. Articulate radicals and militants struggled to create the CIO, but the decade also provided opportunities for thousands of rank-and-file leaders in dozens of unions—stalwarts and rascals, idealists and cynics—to build careers in the labor movement. The surge spanned a wide variety of contexts, both geographical and industrial. We do no disservice to the heroes of the Overpass and the River Rouge, the martyrs of Gastonia and South Chicago, and the leaders of the CIO and its great industrial unions if we insist that we must know more about this seminal decade than their stories alone yield.

The New Deal years were decisive for John P. Burke's union,

the International Brotherhood of Pulp, Sulphite, and Paper Mill Workers of North America. It had declined to fewer than 4,000 members in 1933, many of them Canadians, but it listed more than 60,000 by the beginning of World War II. Though its revival was not as spectacular as that of the United Mine Workers, and its membership gains were not as dramatic as those of some new CIO unions, the Brotherhood did grasp unique opportunities to establish itself in the important pulp and paper industry. Its gains reflected the restlessness and union-mindedness of pulp and paper workers, the stimulation of New Deal legislation, and the commitment to organizing the industry that Brotherhood leaders retained even through the difficult twenties and early thirties.

The union experienced not only rebirth but also decisive change in its composition and character. In the early 1930s, the membership centered on old pulp, paper, and newsprint locals in Canada and in the American Northeast. In some places only the willingness of friendly employers to enforce union-shop contracts kept the Brotherhood in operation. But by 1941 the union had expanded enormously. It had organized the Pacific Northwest and the South, two major areas of new production, and had resuscitated older locals in its areas of traditional strength. Moreover, it had begun for the first time in its history to exercise jurisdictional claims over thousands of workers in the rapidly expanding converted paper industry. Thus in 1933 the typical union member was producing pulp or newsprint in a northern mill town, such as Carthage, New York, or Madison, Maine. In 1941, he (or she) was more likely to be performing tasks in Panama City, Florida, or in Longview, Washington. The typical member might equally well have been a woman manufacturing corrugated containers in Akron, folding boxes in Philadelphia, or paper plates, straws, and novelties in New York City. By 1941, a union that had once been Anglo-American, leavened by French Canadian membership on both sides of the border, had become ethnically diverse, enrolling blacks, Portuguese, Italians, and Jews in the increasingly important urban locals.[10]

The Brotherhood did not have a single dramatic organizing cam-

10. For the pre-1933 history of the Brotherhood, see Voelker, "History of the International Brotherhood"; Brotslaw, "Trade Unionism," 111–32; Cernek, "Beyond the Return to Normalcy"; and Graham, *Paper Rebellion*, 1–14.

paign such as the one that transformed the Mine Workers in 1933. Nor did it undergo a great formative strike, such as the UAW's victory in Flint. Its opponents in the field were in some locales as likely to be other unions scrambling for jurisdictional rights in a growing industrial sector as to be implacable corporate executives, although the union encountered both. The paper industry did not send forth the kinds of working-class radicals, troubadours, and proletarian intellectuals who found the longshoring, auto, steel, and rubber industries so hospitable.

Indeed, its leadership in the 1930s combined an array of disparate characteristics. Loyal to the AFL, these spokesmen sympathized with and sought to practice industrial unionism. Though skeptical of the steadfastness and solidarity of paper workers, they pressed on doggedly with organizing. Unstirred by the pageants, mass strikes, and aggressive rhetoric of the period, they clung to their own distinctive convictions regarding the sanctity of labor, the righteousness of organization, and the honor of struggle. They were socialists, but they were also voluntarists frequently skeptical of the federal government's activities in the field of labor relations. At the same time, they eagerly exploited the changes in labor law to promote their cause. All of these leaders spoke less of revolt than of honoring contracts and persuading employers, as well as workers, that benefits would flow from organization. Some of these hard-working, honest, provincial, and cautious men spent their entire adult lives guiding their union through a stormy period. In their eyes, the times often presented as much of a threat to the union's traditions as an opportunity for expansion, growth, and power.

The Pulp, Sulphite, and Paper Mill Workers remained affiliated with the AFL throughout the 1930s. In its semiindustrial structure, its at least nominally socialist traditions, and its broadly defined jurisdiction, it differed from the unions that dominated the Federation. Its efforts to organize one of the nation's leading industries remain largely unchronicled. It sought to survive and grow as part of a labor movement riven by factionalism and conflict, responding to industrial and unionist circumstances beyond the previous experiences of its leadership. Its involvement with workers in diverse geographic and industrial settings reveals much about the nature of the labor uprising of the 1930s and about the AFL's response to the activism of workers and to the CIO challenge.[11]

Three central factors explain the Brotherhood's resurgence in the 1930s. The first is the sheer fact of the union's survival through years of defeat and decline — its ability to cling to its broad jurisdiction and to remain capable of exploiting opportunities to extend its organization. The second is the labor legislation passed by Congress, notably Section 7(a) of the National Industrial Recovery Act (NIRA) and the National Labor Relations Act of 1935. The third is the challenge presented by the creation and activities of the Congress of Industrial Organization. Both workers and employers were central. The workers themselves, of course, had to be willing to support the union. Influential employers also helped make the climate favorable for unionism. In some places, arrogant and hostile employers created situations that the union eagerly seized upon; elsewhere, corporate executives openly welcomed members of the Brotherhood, viewing acceptance either as an adjunct to modernized personnel policies or as a hedge against the aggressive CIO.

The Pulp, Sulphite, and Paper Mill Workers had been chartered by the American Federation of Labor in 1909. The organization gradually extended its jurisdiction to include workers in converting plants as well as in pulp and paper mills; the Brotherhood experienced both success and galling defeat in the years before 1933. Gains in the 1910s dissolved in a series of disastrous strikes in the early 1920s. The union discontinued its hopeless walkouts in 1926 in time to suffer the crushing effects of the depression. By 1932, it was barely alive, kept afloat in good part by the faithful support of Canadian workers and by union shop contracts in a few American mills. Rank-and-file members castigated the leadership for the defeats of the twenties, while the union's officers grew increasingly skeptical of the willingness of pulp and paper workers to make sacrifices on behalf of the union.

Yet the organization remained alive. Its officers, for all their discouragement, kept plugging away in New England, New York State, and Wisconsin. "There is not much doing in the organiz-

---

11. Brotslaw, "Trade Unionism," 133–46; Graham, *Paper Rebellion*, 14–36; Zieger, "Limits of Militancy," 638–57; and Zieger, "Oldtimers and Newcomers," 188–201.

ing line at the present time," observed President Burke in September 1930, "because nothing much can be done with the mills running so poorly."[12] Still, the organization kept a few men in the field, even when Burke had to suspend publication of its magazine and all but disband the office staff. A core of loyal workers continued to support the organization. Moreover, the people in office remained determined to do more than maintain an organization with sufficient income to service a few locals and to provide key staff with jobs—to do more than had been done by the Amalgamated Association of Iron, Steel, and Tin Workers, another AFL union with wide jurisdiction. Even in the darkest days, the Brotherhood's leaders struggled to preserve broad organizing rights despite the depredations of the craft unions, which were hungry during the depression for every last member. As Burke remarked in more prosperous times, "We haven't an industrial charter from the A.F. of L. Of course, it is semi-industrial, and the reason we are so much of an industrial union is because we have gone out and organized on that basis. Necessity has compelled us to organize that way."[13]

As a result of such survival strategies, when Congress passed the National Industrial Recovery Act in June 1933, an active, if impoverished, organization existed in the pulp and paper industry into which new unionists could be channeled. Burke immediately understood the importance of Section 7(a) as a stimulus to organizing, although he later sometimes understated it lest the union appear to have been a passive beneficiary. "The people of the United States," he said in March 1935, "during the summer and fall months of 1933 were stirred into a sort of religious enthusiasm for the N.R.A."[14] Throughout the country, in areas in which Brotherhood representatives did not even suspect the presence of workers falling under the organization's jurisdiction, recruits in the pulp, paper, and converting industry, eager to better their lot and to support the Blue Eagle program, surged into new locals. Ninety new locals arose, at least seventy of which were still functioning in 1935. Membership rose to more than 13,500 by the

12. Burke to A.C. Krueger, July 25, 1930, Reel 1.
13. Brotherhood, *Report of Proceedings . . . 17th Convention* (1937), 80; Burke to R.C. Henke, June 11, 1930, Reel 1.
14. Brotherhood, *Report of Proceedings . . . 16th Convention* (1935), 2.

end of 1934, compared with the monthly average of 5,700 in 1933 and fewer than 4,000 in 1932.[15]

This surge of membership survived the inadequacies of the legislation that had helped to trigger it. It soon became clear that the guarantees of Section 7(a) could not be enforced. Employers found means of harassing, enticing, and browbeating workers. The union often could not effectively defend or service thousands of the new recruits. Still, while the organization suffered reverses after the initial wave of enthusiasm, and while some workers grew disgusted with its inability to defend activists, Section 7(a) remained crucial. It replenished the union's coffers, it attracted new organizers, it enabled the organization to stake out far-flung jurisdictional claims that the Brotherhood could later exploit, and it boosted the morale of both the leadership and the rank-and-file members. The National Labor Relations Act of 1935, although less spectacular in effect, was critically important in the union's largely successful efforts to sign up major corporations, to frustrate obdurate employers, and to stave off the CIO challenge.

Indeed, by the late 1930s the CIO threat had become serious. Brotherhood leaders continually railed against the rival federation, even though they had supported the industrial union position within the AFL and had opposed the ouster of the CIO unions. Still, despite an occasional defeat by the minions of John L. Lewis, the union clearly benefited from the presence of the CIO. Cautious workers, believing organization inevitable, rushed to sign Brotherhood cards to avoid the "radical" alternative. More important, employers, even traditional foes of unionism, saw the Pulp, Sulphite, and Paper Mill Workers as the lesser of the evils. Many an organizing campaign in the late thirties and early forties was facilitated by friendly cooperation in the company offices. It was no coincidence that 1937, the first full year of CIO operations, was, as Burke proudly told his field representatives, a year of "remarkable accomplishments" for the Brotherhood.[16] After a pause caused by the sharp recession of 1937–39, the organization resumed its rapid growth in 1940–41, spurred by the increasing activity of a CIO paper workers' union and by the forays of United Mine Workers' District 50 into the field.

15. Ibid., 25.
16. Burke to organizers, Dec. 22, 1937, Reel 1.

In responding to the challenges and opportunities of the decade, the International Brotherhood of Pulp, Sulphite, and Paper Mill Workers shared similarities with other contemporary labor organizations. Like the Amalgamated Meatcutters and Butcher Workmen, it promoted industrial organization within the AFL. Like the new CIO organizations in auto, rubber, textiles, and other mass production industries, it claimed to speak for the unskilled. Like unions in oil and chemicals, it centered on a technologically advanced, sophisticated, and rapidly expanding industry deeply enmeshed in the consumer culture of modern America. On the other hand, like the unions in logging, sawmilling, and mining, it was often rooted in remote locales, and, again like those industries, the pulp and paper industry had known sharp labor conflict over the years. Like textiles and apparel, the important converting sector of the industry employed widely diverse ethnic groups and thousands of female operatives, all the while also experiencing rapid geographical diffusion.[17]

In the final analysis, however, the pulp, paper, and converted paper industry was sui generis, as was the Brotherhood. The revival of the union in the 1930s resulted from the dynamic combination of a long tradition of unionism with new government initiatives and a powerful rival unionism. The union's successes in the 1930s, though lacking the dramatic symbolism of the victories at Flint, on the San Francisco waterfront, and in the streets and barricades of Minneapolis, formed an integral part of that fascinating decade.

17. Galenson, *CIO Challenge*, 266–74 (rubber), 300–41 (textiles and apparel), 412–21 (oil); Christie, *Empire in Wood*, 287–300; Jensen, *Lumber and Labor*, 152–214ff.; and Brody, *Butcher Workmen*, 128–204.

2

The Pulp and
Paper Industry

IN THE 1930S THE PULP, PAPER, AND CONVERTED PAPER INDUSTRY was diverse, multicentered, and expanding. Its distinctive character shaped the growth and development of the International Brotherhood of Pulp, Sulphite, and Paper Mill Workers. The industry's two parts differed vastly. The primary pulp and paper sector was highly integrated, consolidated, and heavily capitalized; the converted paper sector was sharply competitive, locally centered, and oriented toward specialty markets. The union was thus obliged to operate in a wide variety of locales and industrial environments. The growth and promise of the industry as a whole offered the organization great opportunities to expand its membership and to fulfill long-standing ambitions. At the same time, the need to deal effectively with vertically integrated national corporations and also to negotiate on an individual basis with family-owned paper-box shops in dozens of cities taxed the ingenuity and stamina of the Brotherhood's representatives. The contrasts held both promise and challenge.

In the thirties, the pulp and paper industry combined strength and potential for vast expansion with dismaying instability. All indexes pointed to a growing demand for its products. Typical citizens found themselves ever more dependent on paper at home, on the job, and at leisure. They were inundated by printed matter, from the morning newspaper to the daily tide of order forms and mimeographed material associated with an increasingly bureaucratic society. Paperboard cartons contained a widening array of food products. Disposable personal care items and other paper products proliferated, many of them supplanting items formerly made of cloth, glass, wood, or metal. Producers, shippers, and builders turned to reinforced and treated paperboard, corrugated containers, sacks, construction materials, and boxes, finding them cheaper, more readily available, and easier to handle than the more traditional wood, glass, and metal. Per capita consumption of paper, which stood at 100 pounds in 1917, grew throughout the 1920s. Despite the setback of the early depression, by 1940 the average American used 265 pounds of paper annually.[1]

Since the emergence of a modern paper industry late in the nineteenth century, production and use of the various grades of paper had shown changing patterns. Until the eve of World War I, major paper producers sought primarily to meet the expanding demand for newsprint, with paper for wrapping and book production occupying an important second place. Early in the twentieth century, developments in the chemistry and technology of papermaking, together with changes in marketing, consuming, and shipping, altered the mix of paper issuing from American mills. The 1913 tariff entered Canadian newsprint on the free list, a victory for American publishers but a defeat for International Paper, Great Northern, and other large newsprint producers. Although Maine and New York State continued to turn out record quantities of newsprint for another decade, American production of this grade of paper declined thereafter, as inexpensive Canadian and other foreign supplies dominated the market.[2]

American paper makers soon found outlets for other grades.

1. Brotslaw, "Trade Unionism," 19; John A. Guthrie, *Economics of Pulp and Paper*, 66; James D. Studley, *United States Pulp and Paper Industry*, U.S. Department of Commerce, Bureau of Foreign and Domestic Commerce, Trade Promotion Series 182 (Washington: U.S. Government Printing Office, 1938), 15–22 and passim; "Economics of Paper," *Fortune* 16 (Oct. 1937), 111–17, 178, 182, 184, and 186; "Fifteen Paper Companies," *Fortune* 16 (Nov. 1937), 192–99; and "International Paper and Power," *Fortune* 16 (Dec. 1937), 131–36, 226, 229, and 230–32.

2. Ellis, *Newsprint*, 69–89 and passim. Changing patterns of production and technology in the industry are best followed in Smith, *History of Papermaking*.

Production of book, wrapping, and fine papers doubled in the quarter century after 1914, providing work for the great machines that had produced newsprint. The most dramatic advances, however, involved new consumer products. Per capita consumption of paperboard—the thin "cardboard" packaging that consumers most commonly encountered in containers for cereal, razor blades or, with a waxed coating, milk—grew from thirty-five pounds in 1917 to almost one hundred in 1940. "Paper drinking cups, sundae cups, soufflée cups, and containers for cottage cheese, peanut butter, and ice cream are known to everyone," a Department of Commerce study observed in 1937. The paperboard milk container, "a somewhat recent development," was rapidly supplanting glass bottles on American doorsteps and in dairy cases. "Over 1,000,000 are used daily in over 100 cities," the report noted. Also increasing rapidly was the production and use of tissue paper for health care and personal use. Per capita consumption jumped to more than twelve pounds by 1940, a sixfold increase over the 1914 figure. In addition, Interstate Commerce Commission rulings in the 1900s and 1910s extended the use of fiberboard and corrugated shipping containers in rail and truck transport, boosting production of boxes and heavy containers to record levels in the 1920s and 1930s.[3]

This proliferating array of consumer and semiconsumer goods kept more than 700 paper mills in operation, even in the early 1930s. Nonetheless, the depression caused hardship in the paper-producing towns and cities. Always prone to cycles of overexpansion and price cutting, paper manufacturers suffered sharp setbacks after 1929. By 1932, net sales had dropped 44 percent from the record high of $1.7 billion in 1929. Although general deflation accounted for some of the decline, production decreased by 28 percent in the same period, and 1932 found manufacturers operating at only 58 percent of capacity. Per capita consumption, which had risen steadily since 1914, plunged one-third between 1928 and

3. The words quoted are in Studley, *United States Pulp and Paper Industry*, 14, 17. Per capita consumption figures appear in Guthrie, *Economics of Pulp and Paper*, 42, 51–54, 63, and 66. Material on Interstate Commerce Commission rulings may be found in "Labor Cost of Production and Wages and Hours of Labor in the Paper Box-Board Industry," Bureau of Labor Statistics, *Bulletin* 407 (1926), 50, and in Bettendorf, *Paperboard and Paperboard Containers*, 73, 77.

1933. Total employment in primary pulp and paper fell from more than 128,000 in 1929 to 108,000 by 1932, a drop of 16 percent.[4]

Even so, the decline in paper was less severe than the setbacks that other major industrial groups experienced. Moreover, pulp and paper recovered rapidly in the mid-1930s. By 1936–37, sales surpassed the predepression levels; in 1937 the industry performed — temporarily, as it turned out — at an all-time high 89 percent of a considerably expanded capacity. In late 1937, almost 138,000 workers toiled in the pulp and paper mills, 7.6 percent more than had done so in 1929.[5]

By the late 1930s, when it was on the verge of major expansion, the paper industry ranked tenth in the country in terms of its products' value. Thirty-eight states contained at least one papermaking facility. Paper, *Fortune* magazine observed, "is constantly being substituted for other materials that are more expensive." Rapid expansion into the South and the Pacific Northwest, the "endless diversity" of paper products, and a strong tradition of associationalism gave the industry a reputation as a modern, dynamic field with a profitable future. In 1941, primary paper production exceeded 17 million tons, a figure 60 percent above 1929 levels, while for the first time sales of all categories of pulp and paper surpassed the $2 billion mark.[6]

4. Per capita consumption figures are in Guthrie, *Economics of Pulp and Paper*, 66; sales figures are in Harry Edward Graham, "History of the Formation of the Western Pulp and Paper Workers" (Ph.D. diss., Univ. of Wisconsin, 1967), 313; figures on capacity and production are in "Economics of Paper," 111, and in Brotslaw, "Trade Unionism," 39. Because of taxonomic confusion and resulting duplication of enumerations, employment figures in pulp and paper are approximations. Those used here are drawn from U.S. Department of Commerce, Bureau of the Census, *Sixteenth Census of the United States: 1940, Manufactures, 1939*, vol. II, pt. 1 (Washington: U.S. Government Printing Office, 1942), 632–33, 649. But see also Smith, *History of Papermaking*, 409, and W. Rupert Maclauren, "Wages and Profits in the Paper Industry, 1929–1939," *Quarterly Journal of Economics* 58 (Nov. 1943–Aug. 1944), 99.

5. "Economics of Paper," 111; Graham, "Formation," 313. For labor force, see *Census . . . Manufactures*, 633; for comparisons with other industries' performance, see Brotslaw, "Trade Unionism," 53–54, and R.A. Lawrence, "Preliminary Report on the Paper and Pulp Industry," Records of the National Recovery Administration (RG 9), Records of Divisions, Records of Research and Planning Division (National Archives), 14–15, 38.

6. The quotation appears in "Fifteen Paper Companies," 199. On associationalism, see Lawrence, "Preliminary Report," 1, 5. On movement into the South

By the 1930s, paper producers had made substantial progress toward vertical integration. Large concerns often controlled the timberlands and sawmills that provided the logs and chips from which pulp was processed. Less often, the company controlled transportation facilities to convey the wood or sawmill residue to its pulp mills. Increasingly, pulp making and paper making—technologically distinct operations—were concentrated in the same facility. Upon arrival at the pulp-making site, the logs were debarked, cut, chipped, and cooked in a chemical mixture under pressure and were beaten and mixed to produce liquid pulp. In the newer facilities, the pulp was then carried within the same complex to the paper-making equipment to begin its progress through the fourdrinier or the calender, the basic instruments of paper making. Some ultramodern paper-making facilities then further processed the machines' product, treating the paper as raw material in the fashioning of wrapping, sacking, construction materials, tissue, or other finished goods.[7]

In practice, however, the industry rarely displayed such thorough vertical integration. No company could control sufficient timberlands to keep itself supplied with wood from its own holdings; woodcutting remained decentralized, with even multiplant corporations relying heavily on small independent contractors, farmers, and woodsmen. Integration of shipping facilities was rare. Pulp and paper manufacturers increasingly relied first on the railroads and later on motor trucking for supplies of wood, chemicals, mixing, coating, and finishing materials.

Even the manufacturing process itself was only partially integrated. Increasingly, pulp and paper were produced in the same facility; in 1936–37, about 80 percent of the volume of paper came from combined pulp-paper mills. Even so, at least half of the pa-

---

and Pacific Northwest, see Smith, *History of Papermaking,* 364–72 and 391–442, and Guthrie, *Economics of Pulp and Paper,* 6–11, 45–50. For sales, Graham, "Formation," 313; for production figures, Brotslaw, "Trade Unionism," 39; for ranking, Studley, *United States Pulp and Paper,* 9, and U.S. Department of Commerce, Bureau of the Census, *Historical Statistics of the United States from Colonial Times to 1970: Bicentennial Edition,* pt. 2 (Washington: U.S. Government Printing Office, 1975), 669–80.

7. For descriptions of the technology of the pulp and paper industry, see Smith, *History of Papermaking;* Kenneth W. Britt, ed., *Handbook of Pulp and Paper Technology,* 2d ed. (New York: Van Nostrand Reinhold, 1970).

per mills in operation did not produce their own pulp, purchasing it either through contract or on the open market from pulp manufacturers. Integrated conversion of paper into finished products was even less developed in the mid-1930s. Thus although the industry had some of the characteristics of continuous-flow enterprise, in important respects it remained divided into discrete operations. Pulp and paper making became increasingly concentrated in adjacent facilities, but the industry tended toward diffuseness at either end of the raw-material-to-consumer-product spectrum.[8]

At the heart of the integration that did exist was the basic mechanical instrument, the large paper-making machine, usually of the fourdrinier variety. Invented in 1803, it retained the same basic design 130 years later, incorporating improvements, electrical power, and increases in size and speed within a framework that its inventor, Nicholas Louis Robert, would have recognized. By the mid-1930s, machines more than 150 feet in length and 70 feet in width, costing more than $500,000, had become common. The fourdrinier dominated the entire process of paper making, receiving liquid pulp at its "wet" end and producing finished paper, usually in huge rolls, at its "dry" end. Increases in size and speed continually reduced the ratio of labor costs to production costs. Indeed, the perfection of machinery and paper chemistry permitted the South and the Pacific Northwest to challenge older sections for supremacy while paying wages that were above regional norms (in the case of the South) or were above industry norms.[9] The fourdrinier machine employed an elaborate series of belts, screens, and felt rollers to process the pulp, which arrived in liquid form or was reliquefied from transported pulp, into paper. The fourdrinier held many advantages, for it was virtually indestructible, easily modified to accommodate chemical and mechanical innovations, and quite versatile, permitting the manufacturer to vary grades and types of paper by adjusting the chemical mix and the screening and rolling operations.

The versatility and durability of the fourdrinier also contributed to the industry's unique blend of corporate consolidation

8. "Economics of Paper," 182; Lawrence, "Preliminary Report," 8–9, 11–12; Brotslaw, "Trade Unionism," 16–17, 28–29.

9. Smith, *History of Papermaking*, 30; Maclauren, "Wages and Profits," 198–99; "International Paper and Power," 229–30.

and rugged entrepreneurial individualism. In the 1880s and 1890s, combines led by International Paper and the Great Northern Paper Company had merged smaller companies into large newsprint-producing corporations with direct ties to major newspapers. The 1913 tariff forced these companies to expand into Canadian holdings, which International Paper did through its subsidiary the Canadian International Paper Company, and to adapt machinery to other grades. Meanwhile, the perfection of the sulphate process of pulp production, which for the first time made the highly resinous southern pine commercially pulpable, encouraged expansion into the American South, a development that International Paper also pioneered through another subsidiary, Southern Kraft.[10] Although the nature of paper making precluded the kind of highly concentrated market domination that was developing in the automobile industry, the need to invest not only in the expensive fourdrinier machines but also in timberlands, water supplies, and transportation facilities ensured that these large firms would gain an expanding share of the growing market for paper goods.

Yet the number of individual firms producing paper remained large. Throughout the thirties, more than 700 companies remained in the field. Many single-machine specialty operations continued, dominating a local market by producing a particular grade of paper or paperboard needed by a local industry. The adaptability of the fourdrinier allowed formerly general-purpose companies, now undercut by industry giants, to shift grades and to enter new product areas. For example, the innovative leadership of D.C. Everest of the Marathon Paper Company in Wausau, Wisconsin, prodded that company to move into specialty papers, such as waxed wrappings for chewing gum, bread, ice cream, and other consumer commodities when central Wisconsin's timberlands suffered overcutting and larger firms took over Marathon's newsprint markets.[11]

Shrewd management and wise utilization of engineering expertise, together with careful selection of properties, also permitted

10. Smith, *History of Papermaking*, 170–76; 198–206, and 391–441; Bettendorf, *Paperboard*, 31.

11. "Fifteen Paper Companies," 192; Smith, *History of Papermaking*, 358, 374, and 445; George A. Prochazka, Jr., "The Pulp and Paper Industry of the United States," revised Dec. 1934, NRA Records, Records Maintained by the General Files Unit, Consolidated Reference Materials, Special Research and Planning Reports and Memoranda.

the maintenance and expansion of family enterprises. The West Virginia Pulp and Paper Company exemplified this pattern. Established in 1888 as the Piedmont Pulp and Paper Company by William Luke, a Scots immigrant, the firm acquired coal and timber properties in Virginia, West Virginia, and Maryland and took over established mills in New York and Pennsylvania. By 1910, its profits topped a million dollars a year, and in the 1930s it ranked as one of the larger firms in the industry while retaining its family ownership and management. Throughout the depression, West Virginia continued to pay dividends, consolidating its position as a leader in the exploitation of southern pinelands.[12]

The distinctive technologies and resource requirements of paper making made it a heavy employer of engineering and scientific talent. In even a family-run company such as West Virginia Pulp and Paper, company officials not only had to oversee and maintain the heavily mechanized processes of pulp and paper making but also had to concern themselves with the acquisition and exploitation of extensive timberlands and coalfields. Technical requirements were sophisticated, calling for the skills of chemists, mechanical engineers, foresters, and water control experts.[13]

At the same time, the industry exhibited a certain old-fashioned individualism. Actual operations began in the remote woodlands and streams, and major production facilities lay not in Pittsburgh or Detroit but in small mill towns and even villages close to water and wood supplies. Thus places with populations below 100,000 housed more than 90 percent of the mills. "The industry is essentially rural in character," noted a government report in the mid-1930s, for "63 percent of all mills are located in towns and villages of 10,000 people or less," while more than a third of all mills stood in villages and hamlets with fewer than 2,500 residents. For at least 137 American communities, "a pulp and paper mill is the chief source of employment." Nor did the individual mills employ large numbers of workers. Paperboard mills, which accounted for about half of total primary paper production, averaged 120 employees. Only a few mills contained as many as 500 workers,

---

12. Smith, *History of Papermaking,* 395–400, 457–58, and 474; "Fifteen Paper Companies," 190–92.

13. Studley, *United States Pulp and Paper,* 13.

while more than 40 percent of the mills included in a 1939 survey employed fewer than 100 people.[14]

Despite its generally happy prospects, the industry suffered from a certain notoriety in business and financial circles. A series of articles in *Fortune* magazine in 1937 referred to the "manic-depressive cycle" of production and to the "cockeyed set of economics." The huge size, immobility, and heavy capitalization associated with the fourdrinier encouraged overproduction. In addition, paper making depended on ready supplies of wood. As the forests receded in New England, New York, and the Great Lakes region, paper makers sought new sources. Large concerns with access to credit moved rapidly into the South, while smaller firms often found themselves with inadequate access to cheap pulp or pulpwood and with heavy costs. Still, once it had been built and installed, the durable fourdrinier tended to remain in operation; manufacturers adapted to changing local market requirements by shifting grades of paper. Even bankruptcy often failed to shut down the huge machines, since receivers found the alternative of producing paper, almost regardless of price, less destructive of their wards' interests than the alternative, the dismantling of the sturdy machines for sale as replacement parts. As a result, the industry tended to retain inefficient older production areas even while it expanded into heavily forested regions. The ever-rising demand for paper products forestalled a complete boom-and-bust cycle, but overproduction throughout the 1920s sustained the industry's reputation as "mad," in the words of *Fortune*'s reporter, "with one of the shortest and most violent cycles."[15]

Potential investors found papermaking a combination of risks and enticements. Rising demand and almost universal predictions of a glowing future were encouraging, as was an overall reputa-

14. [H.J. Martin], "History of the Code of Fair Competition for the Paper and Pulp Industry . . . ," NRA Records, Records of Divisions, Division of Review, Code 120, pt. 1, p. 2; CIO Staff Memorandum, c. February 3, 1944, AFL-CIO Records, AFL-CIO Library, Washington, Reel "CIO National 2": File: "Paper Workers of America," summarizing a variety of government reports on the paper industry dating from the 1930s; on paperboard characteristics, see H.E. Riley, "Earnings and Hours in the Paperboard Industry," Bureau of Labor Statistics *Bulletin* 692 (1941), 1–4. See also Maclauren, "Wages and Profits," 200–201.

15. "Economics of Paper," 111; Brotslaw, "Trade Unionism," 51–54.

tion for high levels of managerial competence and technical expertise. More discouraging to investors were the periodic gluts of paper, together with recourse to "spot" purchasing practices that made prices unpredictable in the short run. Similarly unattractive was the industry's reputation for an extremely low rate of capital turnover due to heavy investment requirements; the low rate of return characteristic of utilities was thus coupled with a short-range element of risk. In the long run, these trends strengthened the hand of the emerging corporate giants such as International Paper. Such firms could better liquidate older operations and could expand into the profitable southern field while tapping the eastern capital markets and using growing market power to establish prices and hence to stabilize conditions of production. Still, by the late 1930s these processes were only partially completed, and in general the industry continued to combine unsettling short-range fluctuations with healthy long-range prospects.

The 120,000 or more workers who toiled in the nation's pulp and paper mills were of course affected by the industry's vagaries most directly. The industry employed a predominantly male labor force. The workers were sharply differentiated according to skill. In the pulp mills, which were fewer in number by two-thirds and were more and more frequently adjacent to paper-making facilities, skilled labor was largely confined to the beater room, where wood chips were mixed with chemicals and were cooked under pressure to yield liquid pulp. In the paper mills, the crews that ran the fourdriniers, together with skilled craftsmen who maintained the machinery and operated the powerhouses and electrical generating equipment, formed the solid core of skilled workers. Skilled labor composed about 20 percent of the labor force in the primary pulp and paper industry.

About a third of the workers were semiskilled. In the pulp mills, the workers who performed the physical tasks associated with loading the huge digesters, handling the acid containers, dumping the alum, clay, and other additives, and unloading the vats represented the bulk of the labor force within the beater rooms. In the paper mills, semiskilled workers included the support crews for the head tour boss, or machine tender, and his chief assistant, the back tender, together with oilers, employees who cleaned the drying and pressing felts, cutters, and trimmers. All of the skilled workers, including the boss beatermen in the pulp mills, fell under the jurisdiction of the International Brotherhood of Paper Makers, the partner and sometimes rival of the Pulp, Sulphite, and Paper Mill Workers; electricians, machinists, and carpenters belonged, as was the case in many mills, to the various craft unions.[16]

The remainder of the semiskilled operatives and all unskilled workers belonged to the Pulp, Sulphite, and Paper Mill Workers' jurisdiction. A wide variety of laborers, including those who cut wood in the timberlands and those who handled, stored, and transported the finished rolls of paper, were unskilled. Yard workers, wood handlers, most of the common labor in the beater rooms, unskilled maintenance help, workers on shipping docks, and bundlers, wrappers, and all miscellaneous labor came into the union's orbit. In the words of its jurisdictional grant from the AFL in 1909, the Pulp, Sulphite, and Paper Mill Workers laid legitimate claim to "grinder men, wood loaders—both inside and outside of the mill—all screen men, floor sweepers, oilers, press tenders, decker men, wet machine tenders, digester cooks, and all help employed on or around the digesters; acid makers and all help employed in and around the acid-making plants."[17]

Wage rates closely followed skill gradations. Machine tenders and their crews were highly identified with their jobs and were always in demand. In the half century before the Great Depression, these workers established and maintained high wages, usually pegging their scales to the rates enjoyed by the skilled craftsmen who repaired the fourdriniers and manned the power plants. Thus, for example, a 1925 survey placed wages for machine tenders and beater engineers at 79.9 cents and 66.9 cents per hour, respectively, at a time when the average hourly wage for manufacturing workers stood at 54 cents and that for workers in nondurable goods production somewhat lower. In contrast, the average wage for beater helpers that year was only 43.5 cents. An industry survey in 1934 revealed this disparity even more graphically. Machine tenders earned about 83 cents an hour, while beater-

16. "Labor Costs of Production," 86–89; Riley, "Earnings and Hours," 14–15, 30; MacDonald, "Pulp and Paper," 103–18; Brotherhood, *Constitution and By-Laws* (1933), 3–4. There are excellent photographs and diagrams of the processes involved in work in pulp and paper mills in Smith, *History of Papermaking*, Bettendorf, *Paperboard*, and Studley, *United States Pulp and Paper*.

17. Brotherhood, *Constitution and By-Laws*, 3–4; Riley, "Earnings and Hours."

men commanded only about 45 cents. Woodyard, janitorial, and paper-handling workers earned about 3 cents an hour less than the beatermen. Thus, a differential of nearly 40 cents an hour, or nearly 50 percent, was standard in the industry. Although wage rates in the industry normally fell about 10 percent below the national average for industrial workers, scholarly observers noted that the quasi-rural setting of pulp and paper making lowered the cost of living, while in both stability of employment and resistance to wage cutting the industry compared favorably with other pursuits.[18]

The pulp and paper industry was perhaps too isolated and insufficiently colorful to attract the teams of sociologists, journalists, and other sympathetic observers who studied and restudied the American textile, steel, leather, and automobile workers in the 1910s, 1920s, and 1930s.[19] Still, the snippets of information that government investigators gleaned from occasional surveys suggested a constricted and arduous life for the pulp and paper workers. Shifts of eleven and thirteen hours had been standard until the mid-1920s and persisted in many locales. Prior to the depression, the industry had been phasing out seven-day operations, but in 1933, President Burke reported that because of the depression, "in such cities as Hudson Falls, New York, and International Falls, Minnesota, . . . you . . . hear the roar of the paper machines

18. Bureau of the Census, *Historical Statistics*, pt. 2, p. 170; Riley, "Earnings and Hours," 30; MacDonald, "Pulp and Paper," 110–11; Maclauren, "Wages and Profits," 224–28.

19. See, e.g., Liston Pope, *Millhands and Preachers: A Study of Gastonia* (New Haven: Yale Univ. Press, 1942); Tom Tippett, *When Southern Labor Stirs* (New York: Jonathan Cape and Harrison Smith, 1931); W. Lloyd Warner and J.O. Low, *The Social System of the Modern Factory*, Vol. III: *The Strike: A Social Analysis*, Yankee City (New Haven: Yale Univ. Press, 1947); Robert S. Lynd, and Helen Merrill Lynd, *Middletown: A Study in Contemporary American Culture* (New York: Harcourt, Brace, 1929), and *Middletown in Transition: A Study in Cultural Conflicts* (New York: Harcourt, Brace, 1937); McAlister Coleman, *Men and Coal* (New York: Farrar and Rinehart, 1943); John A. Fitch, *The Steel Workers* (New York: Charities Publications Committee, 1910); Interchurch World Movement, Commission of Inquiry, *Report on the Steel Strike of 1919* (New York: Harcourt, 1920); Charles Rumford Walker, *American City: A Rank-and-File History* (New York: Farrar and Rinehart, 1937); Ruth McKenney, *Industrial Valley* (New York: Harcourt, Brace, 1939); Frank Marquart, *An Auto Worker's Journal: The UAW from Crusade to One-Party Union* (University Park: Pennsylvania State Univ., 1975).

on every Sunday" once again.[20] The mills were noisy, clamorous places and were often hazardous as well. Accident rates surpassed all-industry averages; indeed, some departments regularly reported rates 200 percent and even 400 percent above general norms. Pulp and paper making entailed a great deal of sheer physical labor, work with noxious and dangerous chemicals that required close attention to complex grinding, cutting, and pressing machinery. Splashing and spilling in the acid and cooking rooms took their toll, while the Jordan, fourdrinier, and cylinder machines claimed their quotas of mangled limbs, crushed toes, and severed digits.[21]

Life in the paper-making towns provided more in the way of community recreation and outdoor sports than was typical for workers in other industries. At the same time, air and water pollution—most pervasively the stench that accompanied some pulping processes—eroded the quality of life. In the one published survey of paper workers' families, federal economists in 1935 found that residents of Berlin, New Hampshire, a city of 20,000 and the home of a large paper mill, averaged $1,133 a year in income. Paper making offered few employment opportunities for women, so most families had but one breadwinner. The yearly budget allowed $400 for food, $120 for clothing, $150 for housing, $100 for utilities and fuel, $45 for transportation, $45 for medical care, and $70 for recreation and education. After spending about forty-five cents a week for religious and charitable purposes—Berlin was a church-going, Roman Catholic city, with many French Canadians—the family, if it avoided major disaster, emerged from a year of full employment $25.00 ahead.[22] Union members parted with $12.00 of this amount for their dues.

The converting industry differed from primary pulp and paper

20. John P. Burke, Statement, September 14, 1933, Hearing 405-1-04, Paper and Pulp Industry—Proposed Revisions, NRA Records, Records Maintained by Library Unit, Transcripts of Hearings.

21. Edward R. Jones, "Paper Industry Study," March 5, 1936, 176, NRA Records, Records Maintained by the General Files Unit, Consolidated Reference Materials, Consolidated Files of Typescript Studies Prepared by the Division of Review; Frank S. McElroy and George R. McCormack, "Work Injuries in Pulp and Paper Manufacturing, 1939–49," *Monthly Labor Review* 71 (Sept. 1950), 338–42, and McElroy and McCormack, "Paperboard: Container-Indsutry Work-Injury Rates, 1938–50," *Monthly Labor Review* 73 (Dec. 1951), 675–80.

22. Faith M. Williams and Gertrude Schmidt Weiss, "Money Disbursements

in virtually every respect. Although large multipurpose firms such as the Robert Gair Company had started to consolidate and to integrate converting and primary paper operations, the manufacture of boxes, shipping containers, construction materials, novelties, and other direct consumer goods remained largely decentralized. In 1940, the Wages and Hours Division of the U.S. Department of Labor found more than 2,300 paper converting establishments in the country, with all states and most cities containing facilities. Large companies, such as the Container Corporation, the Robert Gair Company, Union Bag and Paper, Kimberly-Clark, and Scott Paper, operated large-scale enterprises, especially in the bag and tissue fields. Smaller firms supplied paperboard and packaging to national and regional producers of consumer goods, especially foodstuffs. Primary paper producers such as International Paper were beginning to gain entry into the converted field, either by setting up converting operations in the new plants under construction in the South or by absorbing smaller converting operations. Still, as late as the eve of World War II, most converted paper operations were conducted in small, single-owner shops, producing for local agricultural, jewelry, and garment manufacturers. Few converting shops employed as many as 1,000 workers, the industry average being sixty-five. Nonetheless, converting paper into boxes, containers, novelties, and related products was big business, employing about 200,000 workers in 1940 and ringing up sales of more than $1 billion.[23]

"Paper," declared *Fortune* magazine, "really isn't an industry at all, but a group of several disparate industries." Government reports in the thirties suggested a bewildering variety of distinct products in the field. In 1933 and 1934, the National Recovery Administration established at least eighteen codes of fair competition in the paper industry, with separate ones regulating folding paper boxes, set-up paper boxes, shipping containers, bags,

of Wage Earners and Clerical Workers in Thirteen Small Cities, 1933–1935," Bureau of Labor Statistics *Bulletin* 691 (1942), 1, 29, 45, and 112.

23. CIO Staff Memorandum, c. Feb. 3, 1944; Bettendorf, *Paperboard*, 52–60 (folding boxes), 61–93 (corrugated and solid fiber boxes), 94–112 (tubes, drums, etc.); and passim; U.S. Department of Labor, Wage and Hour Division, Press Release, July 16, 1940, copy in American Federation of Labor Papers, State Historical Society of Wisconsin, Series 4, Industry Reference Files, Box 97; *1848–1948: A History of the Wisconsin Paper Industry*, 28, 30, 34–46, 51–52.

food dishes and paper plates, milk bottle caps, drinking straws, "Open Paper Cups and Round Nesting Paper Food Containers," "Fluted Cups, Pan Liners, and Lace Papers," and a half dozen more, in addition to the basic pulp and paper and newsprint codes. At the other end of the decade, the Wages and Hours Division of the Department of Labor issued regulations governing wages and hours for the converted paper industry, which included "folding paper boxes; shipping containers; paper cups; bottle caps and hoods; waxed papers; waterproof paper; glazed and fancy papers; commercial envelopes; tags and labels; gummed papers; carbon papers; towels and toilet tissues; auto panels; photo mountings; die cut specialties; sensitized papers; sandpaper; loose-leaf and blank books, tablets and pads and index cards; stationery; playing cards; book matches."[24]

Sharp differences in labor force and traditions of employment characterized the two distinct sectors of the entire paper industry. Whereas much of primary pulp and paper production took place in rural locales, converted paper manufacturing concentrated in urban places, in close proximity to markets. The labor force of the older sector of the industry remained largely Anglo-American, French Canadian, and, in the Middle West, German American. In the new southern mills, large numbers of blacks performed janitorial and common labor jobs, while the paper and bag mills of the South and Pacific Northwest employed discernible numbers of females, unlike the northeastern factories. Still, compared with the converted paper industry, the primary production sector exhibited a large degree of ethnic and sexual homogeneity. Thus, Jews, Italians, Portuguese, and French Canadians worked in the container factories of the East Coast, while Polish American and German American workers toiled in the converting plants of Chicago and Milwaukee. Throughout the converted paper industry, large numbers of female workers were employed, especially in the low-skill and low-wage sectors, for example in the manufacture of set-up boxes. A 1940 survey found that more

---

24. "Fifteen Paper Companies," 132. The histories of the various codes adopted during the NRA period are contained in NRA Records, Records of Divisions, Division of Review. These "Histories," most of them dated c. March, 1936, briefly describe the manufacturing processes, technologies, labor force and market characteristics, and special features of each of these categories of the converted paper enterprise. See also Wage and Hours Division, Press Release, July 16, 1940.

Logs being prepared for pulping in Millinocket, Maine (*Forest History Society*)

Paper workers toiling at the "dry" end of a middle-sized fourdrinier, c. 1940, in Wisconsin (*Forest History Society*)

UNION MADE
PAPER PRODUCTS

Emblem of the International Brotherhood of Pulp, Sulphite, and Paper Mill Workers (*Brotherhood* Journal; *United Paperworkers' International Union*)

John P. Burke, president-secretary of the Brotherhood, 1917–65, in 1934
(*Brotherhood* Journal; *United Paperworkers' International Union*)

than one-third of the country's 200,000 workers in paper products were women and girls. On the East Coast, wastepaper operations employed some black workers, while union organizers in California found that such low-status jobs were often held, sometimes illegally, by Spanish-speaking workers.[25]

Each segment of the industry had a distinct character. Producers of stationery and related products were concentrated in the Northeast and the lake states, although the production of cheaper tablets and envelopes was more decentralized. Established stationery firms had a long tradition in the industry and stubbornly held out against mass production trends. Stationery manufacturers, for example, remained the only major element in the industry to continue to use substantial amounts of rags, formerly the very stuff of paper production in all grades. With their locations in isolated mill towns, they were closely integrated with primary pulp and paper making and often, as in the case of the S.D. Warren Company of Massachusetts, were once general-purpose manufacturers who specialized in stationery after technological changes and resource depletion rendered them no longer broadly competitive.[26]

In contrast, the boom in sanitary tissue for bathroom and personal use developed following World War I. Firms such as Kimberly-Clark and Scott moved rapidly into this sector, capitalizing on the rapidly expanding consumer market, shifting to almost exclusive concentration on sanitary tissues, and shedding the diverse functions they had performed in the nineteenth century. Economic production of sanitary tissues required tremendous capital investment; the low unit value of the product required that labor costs remain minimal, and constant expansion into mass markets necessitated heavy advertising. Thus the tissue industry was characterized by relatively large firms with at least regional markets. They used advanced technologies and sophisticated marketing and production techniques and needed to keep labor costs to a relatively small percentage of production costs. Tissue and sanitary producers were among the paper industry's innovators

25. Wage and Hours Division, Press Release, July 16, 1940; Henry Segal to John P. Burke, Oct. 2, 1941, Reel 6; Frank C. Barnes, Jr., to Burke, c. Feb. 16, 1937, Reel 1.

26. Smith, *History of Papermaking*, 158–63, 323–24, and 337.

in employee welfare programs and in the promotion of company unionism.[27]

Stationery, tissue paper, and bags were items so close to primary paper production that statisticians sometimes excluded them from the converted paper category. Producers of these commodities were located close to raw materials and fashioned their products directly from the matter emerging from the dry end of the fourdrinier, so that they identified closely with traditional pulp and paper making. Manufacturers of set-up paper boxes stood at the other end of the spectrum. The term "set-up" referred to paper containers, usually hand made, produced to fit the precisely specified needs of particular customers. Since these boxes were cheap, bulky, and often fragile, they had to be produced near the manufacturer of the product that they were designed to contain. A report issued by the Bureau of Labor Statistics in 1937 observed that "the very nature of the industry . . . tends to locate [set-up] box factories in places with low rentals, which are usually one or two floors of old loft buildings, often with poor sanitary facilities . . . [and] inadequate light and ventilation. Cellar workrooms were still common in New York City." Set-up containers "must in most instances be carefully constructed of card or boxboard, covered with various papers, and other materials," noted the National Paper Box Manufacturers' Association. "Many of these boxes contain partitions, slots, divisions, compartments, drawers, etc. There are hundreds of sizes, types, styles and designs."[28]

Although application of power machinery in the set-up industry had begun in the 1890s, it made slow progress. Employers in this slowly declining sector found it wiser to exert pressure on labor costs, which accounted for more than 25 percent of production

---

27. Britt, ed., *Handbook*, 550; *1848–1948*, 28–52; Smith, *History of Papermaking*, 361; Charles N. Glaab and Lawrence H. Larsen, *Factories in the Valley: Neenah-Menasha, 1870–1915* (Madison: State Historical Society, 1969).

28. Bettendorf, *Paperboard*, 33–51; [Victor E. Baril and Frances Jones], comps., "Wages, Hours, and Working Conditions in the Set-Up Paper-Box Industry, 1933, 1934, 1935," Bureau of Labor Statistics *Bulletin* 633 (1937), 72; National Paper Box Manufacturers' Association to Wayne Lyman Morse, Chairman, Industry Committee 14, Converted Paper Products Industry, Wage and Hour Division, Oct. 10, 1940, copy in AFL Papers, State Historical Society of Wisconsin, Series 4, Industry Reference Files, Box 97.

cost, than to invest in expensive labor-saving devices. In 1940 the Paper Box Manufacturer's Association counted 865 shops, employing nearly 40,000 workers, 75 percent of them women. Nearly 400 of these shops were in the northeastern quadrant of the country.[29]

Tissues, personal stationery, and set-up boxes, however, represented only diverse extremes of paper converting. Many of paper's products were so new that neither the industry nor the government could generate reliable statistics. Moreover, some converted paper workers toiled in plants which made containers or bags directly from primary paper produced on the same site. Also, some light manufacturing establishments in apparel and confections had box shops connected to the plant and under the same management. Clearly, no more than a few thousand workers, most of them female, worked in the various branches of the industry on the production of paper cups, straws, fluted liners, and related products. There were other areas, however. Government reports in the late 1930s counted at least 15,500 in the folding paper box industry and more than 60,000 in the cluster of subspecialties that fell under the heading of "paperboard containers and boxes." The same report found about 8,700 envelope makers and more than 11,000 paper bag workers. Throughout the decade, union organizers kept stumbling on specialized segments of the industry employing workers who came within the union's jurisdiction but who had characteristics and conditions about which they were completely ignorant.[30]

Wages in converted paper reflected the diversity of the industry. Common labor rates were comparable to those in the primary sector, but most jobs in converting were unskilled, thus reducing average earnings. A 1937 survey of Ohio workers revealed that while those employed in primary pulp and paper averaged 59.5 cents an hour, corrugated and set-up workers struggled with rates of about 42 cents. On the other hand, folding box workers earned

29. National Paper Box Manufacturers' Association to Wayne Lyman Morse, Oct. 10, 1940.

30. Bureau of the Census, *Census . . . Manufactures, 1939*, 633, 666; History of the Code of Fair Competition for the Fluted Cup, Pan Liner, and Lace Paper Industry (Code 296, Aug. 5, 1936), NRA Records, Records of Divisions, Division of Review; [Victor E. Baril and Frances Jones], comps., "Wages, Hours, and Working Conditions in the Folding-Paper-Box Industry, 1933–1935," Bureau of Labor Statistics *Bulletin* 620 (c. 1937), 54–64.

an amount that approached the basic industry average, in part reflecting the larger percentage of males employed in that specialty[31] as well as the more advanced technology used in production of folding boxes.

Male-female differentials were substantial in all sectors. Racial and regional disparities further complicated the wage structures of both primary and converting trades. At common labor levels, women were penalized four cents an hour, as were black workers in the expanding southern mills. In general, southern workers could expect to earn about 8 percent less than northern workers, although the demand for machine tenders and mechanics became so great in the new mills that rates for these favored workers by the late 1930s surpassed those of their northern and middle western counterparts. Larger machines, the latest technology, and plentiful wood supplies made it possible for Pacific Northwest pulp and paper makers to pay the very highest wages. At the other extreme, some southern converting workers earned as little as fifteen cents an hour, while in the mid-1930s female set-up box workers in New York struggled to raise their wages above thirty cents an hour.[32]

Unionists regularly protested both the regional and sexual disparities as well as the generally low level of wages for common labor. Burke predicted that the sanctioning of lower wages for female labor would result in the loss of jobs for men. He also argued that standard minima prevailing in the industry, particularly in converting plant specialties, made it impossible for workers to achieve an adequate standard of living. He noted in 1933, for example, that the Bureau of Labor Statistics had established a weekly income of thirty-one dollars as necessary for a decent standard of living. Rates for common labor, however, yielded weekly pay

31. Viva Boothe and Sam Arnold, *Prewar, War, and Postwar Earnings, Hours, and Employment of Wage Earners in Ohio Industries, 1935–1946* ([Columbus]: The Bureau of Business Research, Ohio State Univ., 1949), 387–403.

32. [Baril and Jones], comps., "Wages . . . Folding-Paper-Box . . . , 1933–1935," 1–4, 27; [Baril and Jones], comps., "Wages . . . Set-Up Paper-Box, 1933 . . . 1935," 1–3, 43–47; unionists John P. Burke, Herbert W. Sullivan, and Charles Vonie, statements on the various NRA codes in pulp, paper, and converted paper, 1933–34, NRA Records, Records Maintained by the Library Unit, Transcripts of Hearings; MacDonald, "Pulp and Paper," 110–11; Riley, "Earnings and Hours in the Paperboard Industry," 1–30.

of fifteen to eighteen dollars. Throughout the 1930s, one of the union's primary goals was the narrowing of differentials between skilled and unskilled labor in the primary industry and the improvement of base rates in the various parts of the converting industry.[33]

By the 1930s, parts of the converted paper industry had appeared in all states and in most cities. Processors used vast quantities of paper and paperboard for cereal and soap boxes. Parts and appliance manufacturers relied on corrugated boxes for shipping to assemblers and retailers; construction activities spawned manufacturers of paper and paper-based packing, insulating, and covering materials. Mechanization was incomplete. Automatic and semiautomatic scoring, shaping, rolling, and gluing devices had made inroads, particularly in the production of corrugated and folding boxes and of disposable consumer goods. Still, the industry employed more than 200,000 workers by 1940, with growing demand and ever-expanding product lines compensating for technological displacement. From the huge, semiautomated tissue plants astride the rivers of Wisconsin to the ramshackle, crowded, and dirty lofts and cellars of New York and Philadelphia, converted paper workers produced goods increasingly important in the flow of commerce.

So diverse and, in many cases, so new were the activities of the converted paper sector that few data were available regarding health, safety, and working conditions in the 1930s. Set-up box workers, most of them women, performed their repetitive tasks in older, low-rent sections of large cities. New York, metropolitan New Jersey, Philadelphia, and the Boston-Providence complex employed at least half of the industry's total. While power machinery had only sporadically been applied to the industry's operations, manufacturers claimed that "a great number of the plants . . . could install at once automatic machinery which would greatly increase . . . output . . . with *the use of much less labor.*" Because of the low skill levels, the industry tended to retain its female workers, many of whom felt that they could not find alternative employment and compiled lengthy service records. Commenting on the

33. John P. Burke, Statement, Hearing 401-1-01, Paper Bag Manufacturing Industry, Nov. 2, 1933, NRA Records, Records Maintained by the Library Unit, Transcripts of Hearings; MacDonald, "Pulp and Paper," 113-16.

recent history of the trade, in 1942 the business manager of New York's box makers' union reported that 90 percent of the employees were now women, "many of them aged workers," trapped in a declining industry.[34]

Folding paper box workers' experience differed sharply. About three-fourths of these people were men. They performed "light, clean work," making cartons for razor blades, packaging for butter, suit boxes, and other consumer items. Geared to mass production markets, folding paper box manufacturers had mechanized their operations rather substantially. Power presses and other devices which, the Bureau of Labor Statistics noted, "will automatically fold, glue, and deliver 'flat' and ready for shipment from 20,000 to 70,000 small boxes per hour" had eliminated much hand labor. Employees required little training, and as a result, employment was sporadic, almost casual in some cases, with high rates of turnover and little employee identification with the firm or the industry.[35]

Few pulp, paper, or converted paper employers provided significant fringe benefits, recreational facilities, or other emoluments of welfare capitalism. In the primary field, companies did sponsor athletic teams, picnics, and other public events, but they did so more as part of their community obligations than to improve their personnel practices. The set-up industry struggled with too much competition and operated too near bankruptcy to provide even minimal amenities, such as eating areas. In some cases even adequate sanitary facilities were lacking, and the companies hardly thought of sponsoring dances, contests, or ball teams. The situation was similar in the folding paper box trade, where the Bureau of Labor Standards reported in 1936 that "health, social, and athletic activities are negligible," although in some plants a daily ten-minute break was afforded to all females and men who smoked. Few employers in any segment of the industry featured paid vacations, pension plans, sick pay, insurance, or other such benefits.

---

34. National Paper Box Manufacturer's Association (NPBMA) to Morse, Oct. 10, 1940; NPBMA, "Supplementary Brief in Support of a Continuance of the 36 Cent Minimum Wage . . . , Aug. 1, 1942"; A.N. Weinberg, Business Manager, Brotherhood Local 299, to Boris Shishkin, Nov. 20, 1942, all in AFL Papers, State Historical Society of Wisconsin, Series 4, Industry Reference File, Box 97.

35. [Baril and Jones], comps., "Wages . . . Folding-Paper-Box . . . , 1933–1935," 36, 52–53.

Hiring and firing tended to be casual, especially in the converting sector. In primary pulp and paper, larger mills had employment offices, but personnel decisions were heavily influenced by personal acquaintance and family reputation in the smaller mill towns. Large employers in the small pulp and paper towns liked to think of their companies as providing important charitable support, both for individual workers down on their luck and for the labor force as a whole; the executives prided themselves on the industry's ability to keep employment levels and wage rates within 20 percent of 1929 standards despite the depression. But such attitudes expressed themselves in ad hoc gestures, without being part of employees' contractual rights.[36]

Patterns of industrial relations in the entire industry remained unsettled in 1933. On the one hand, some employers in the traditional pulp and paper production areas maintained close associations with the two major unions, the International Brotherhood of Paper Makers and the Pulp, Sulphite, and Paper Mill Workers. Some companies honored union shop contracts even in the depression and thereby earned the unions' gratitude. "A lot of our Maine locals [are] kept alive [only] with a union [shop] agreement," admitted Vice-President Herbert W. Sullivan. Union leaders of this generation had come directly from the pulp and paper mills and prided themselves on their knowledge of its processes, problems, and overall character. They closely and sympathetically followed technical and economic developments and deliberately maintained contact with managers, engineers, and publicists in the trade. They regarded themselves as fellow professionals, very much in league with owners, managers, and technical experts when it came to questions of the overall health of the industry, the need for favorable government policies, and utilization of im-

36. Ibid., 48 and passim; [Baril and Jones], comps., "Wages . . . Set-Up Paper-Box . . . , 1933 . . . 1935," 63–73; Maclauren, "Wages and Profits," 224–28; George Miller, General Manager, West Virginia Pulp and Paper Company, Covington, Virginia, to James W. Towsen, March 19, 1934; Towsen to Miller, March 21, 1934, Westvaco Papers, Department of Manuscripts and University Archives, Cornell University Libraries, Box 39. The correspondence between Miller and Towsen, who directed the company's personnel policies from New York offices in the 1930s, contains many references to employee welfare and recreation programs and includes many issues of the Employees' Protective Association newsletter (Boxes 39–41).

proved technologies. Burke declared in 1933, "You take this industry, as a whole, it is an industry that is friendly to labor. It is not
antagonistic to organized labor."[37]

There was another side to the picture, however. If the pulp and
paper industry was in some respects a model of high technology
—a modern enterprise with tendencies toward cordial labor relations—it also possessed a heritage of conflict. Its strikes and confrontations, which dated to the 1880s, may not have achieved the
epic proportions of Coeur d' Alene, bloody Ludlow, or the smokey
struggles of the steel workers, but they had left their own unique
bitternesses and their lingering wounds among mill people. The
1920s were years of struggle and confrontation, as International
Paper and other large corporations sought to reverse gains that
the unions had made during World War I. International Paper's
attempt in 1921 to impose wage reductions and open-shop stipulations triggered a walkout that lasted in some places until 1926.
"Local unions [were] annihilated in scores of places," lamented
Burke, as workers straggled back into the mills, resentful of the
company, the union, and their fellow workers. Antagonisms lin-
gered. Memories remained vividly personal in the face-to-face relationships that characterized community life in the mill towns.
The strike soured hundreds of workers, because the settlement
on which the union had insisted in 1924 had represented a galling defeat for workers who had remained firm for three years. In
another strike in Holyoke, Massachusetts, in 1923, members of
the Paper Makers had followed their international union's directives and had honored their contract by crossing pulp workers'
picket lines. Seven years later, organizers for the unions in the
city reported that "a couple of Poles at the meeting . . . said that
they did not want to join a union . . . that would make them *scab*
on other workers" and found themselves in opposition to Irish
workers spearheading the organizing drive. In Minnesota, a perplexed local unionist reported in 1933, "We have several . . . that

37. John P. Burke, Statement, Sept. 14, 1933, Hearing 405-1-04, Paper and Pulp
Industry—Proposed Revisions, NRA Records, Records Maintained by the Library
Unit, Transcripts of Hearings; Sullivan to Burke, Dec. 22, 1933, Reel 1; Graham,
Paper Rebellion, 1–14; Voelker, "History of the International Brotherhood"; Cernek, "Beyond the Return to Normalcy"; John H. Malin, "A Historical Sketch
of the Formation and Growth of the International Brotherhood . . . ," Brotherhood *Journal* [18] (Oct. 1934), 11; 19 (Jan. 1935), 13–14; (April 1935), 4–6.

were members of the union in 1921 that went out scabbing"; indeed, the man elected president of Local 97 that year proved to be a man who had scabbed in 1921.[38]

In the glow of labor-management harmony that attended the NRA phase of the New Deal, unionists repeatedly praised the industry for its enlightened labor relations policies. But memories of the past, when employers had imposed the open shop and had enlisted the clergy, the press, and the business establishment against the unions, as well as daily reports from organizers describing threats, blandishments, firings, and harassment of union sympathizers, provided reminders of the industry's mixed attitudes. Louisiana activist Sam Moncrief asserted in August 1933: "There has been about 12 men fired down here on account of this local but we will continue to battle." Similar reports drifted in from Wisconsin, Maine, West Virginia, and most other areas of union activity, reinforcing the opinion of union leaders that however enlightened they found some employers, the struggle for a union would often be long and difficult.[39]

The much more decentralized and diverse converted paper field had little experience in regard to unionism and collective bargaining. Organization was fragmentary and in many subspecialties nonexistent. New York City's paper box makers had tried repeatedly to form a union, the latest effort resulting in a federal labor union that kept a tenuous hold on the city's box shops in the mid-thirties. Employers in the highly mechanized tissue-paper specialty had resorted widely, and successfully, to company unions. The Pulp, Sulphite, and Paper Mill Workers had granted a charter to a group of paper plate and novelty workers in New York City in 1917 but otherwise had little representation in the vast converted paper field. Aside from the New York shops, workers had established no tradition of laborite activism. Burke remarked in 1939 that "unionism in the paper box industry has no particular history."[40]

38. Burke to George Roller, Sept. 29, 1933, Reel 3; Andrew Seim to Burke, Aug. 19, 1933, Reel 1; George Brooks to Burke, Nov. 10, 1930, Reel 1. See Cernek, "Beyond the Return to Normalcy," for a full-length account of the strikes of the 1920s.

39. Burke to Helen Safferstone, March 16, 1933, Reel 3; Sam Moncrief, Corresponding Secretary of Local 143, Hodge, La., to Burke, Aug. 25, 1933, Reel 2.

40. Remarks of A.N. Weinberg, Brotherhood, *Report of Proceedings . . . 20th*

Regardless of the past, however, by the time the depression had entered its fourth year, workers in almost all sectors of the pulp, paper, and converted paper industry had become restive and angry. By 1933, wage cutting had become rampant, particularly in converted paper. "You are playing with dynamite," Burke warned NRA code authorities; even the normally quiescent paper workers, he predicted, would soon be at the center of "strikes and trouble . . . that nobody can stop" if wages and purchasing power continued to decline. Matthew Burns, president of the Paper Makers, noted that workers throughout the industry had grown angry and bitter. Once a wave of strikes starts, he warned, "it spreads very quickly." In October 1933 the usually accommodating Burke advised officials at the NRA hearing on the paper box code that although there had been little labor trouble in the past half century among paper box workers, as observers had frequently noted, many workers had now reached the end of their patience. "I want to make it as clear as I possibly can," he said, "that I have never known a time in all my experience in the labor movement when there has been so much discontent among the working people in our industry as there is today."[41]

The onset of the New Deal brought changes. The vast paper industry was filled with paradoxes, mixing a rosy long-term future, dispiriting fluctuations and uncertainties, ever-accelerating technological and chemical innovation, and highly paternalistic or reactionary labor policies. Converted paper operatives also attempted to accommodate the old and the new. Set-up paper boxes, one of the largest employers in the field, used nineteenth-century hand labor processes and faced heavy competitive pressures, while new and highly mechanized facilities produced with

---

Convention (Oct. 9–13, 1944), 66–72; Burke to Fred G. Hoodwin, Dec. 22, 1939, Reel 7. The union's jurisdiction was expanded in 1922, 1926, and 1935 to embrace virtually all converting workers. Brotherhood leaders believed that these extensions merely formalized implicit jurisdictional rights held from the issuance of the organization's charter in 1909.

41. The quotation that mentions dynamite comes from Burke, Statement, Hearing 401-1-01 (Paper Bag Manufacturing Industry), Nov. 2, 1933, NRA Records, Records Maintained by the General Files Unit, Consolidated Files, Consolidated Files on Industries Governed by Approved Codes, Paper and Pulp Industry. The remarks made at the paper box code hearings occur in Burke, Statement, Hearing 406-03 (Set-Up Paper Box Industry), Oct. 3, 1933, NRA Records . . . , Transcripts of Hearings.

little hand labor mountains of tissues, paper bags, and folding paper boxes.

Employers in the industry ranged from huge corporations such as International Paper, which had numerous plants and thousands of workers, to small city paper box shops employing a few elderly women. In some segments, such as newsprint, an unbroken pattern of labor-management cooperation survived the depression years. In other cases, a past of angry conflict and bitter feelings divided workers and companies. In the large converted paper sector, labor-management relations had settled into no firm pattern as yet.

At first glance a New York City box maker seemed to have little in common with a Pacific Northwest digester cook. Identifiable common interests seemed equally improbable among West Virginia wood handlers, operators of scoring and stitching machines in Akron, and black maintenance and yard men in the rising kraft mills of Southern Kraft. Yet these diverse workers shared two important common ties. These people were all part of the long and varied process of paper production, whether they chopped the trees and shoveled the sawmill chips in the forests and woodyards or packaged and bundled the tons and tons of boxes, plates, containers, and cartons on which the country increasingly depended. All of them fell under the jurisdiction of the International Brotherhood of Pulp, Sulphite, and Paper Mill Workers, which sought throughout the turbulent decade of the 1930s to organize them and to provide the industry in which they spent their working lives with strong, united, conservative trade unionism.

3

John Burke and
the Union

HISTORICAL MEMORY PLAYED A SIGNIFICANT ROLE IN SHAPING the responses of the International Brotherhood of Pulp, Sulphite, and Paper Mill Workers to the challenge of organizing the paper and converted paper industries of the New Deal period. The union's support for industrial forms of organization and its leaders' commitment to organizing the unorganized rendered it readier than many AFL affiliates to seize upon the opportunities afforded by Section 7(a) and by workers' militancy. At the same time, the union's long and difficult path bred a certain caution and distrust of aggressive action. John P. Burke, president-secretary of the Brotherhood since 1917, reflected the ambiguity of his union's past experience as he sought the correct combination of energy and restraint necessary to establish unionism throughout the industry.[1]

The Brotherhood was part of a complex pattern of unionism in pulp and paper making that dated to at least 1884. In that year, a group of skilled machine tenders in Holyoke, Massachusetts,

---

1. Cernek, "Beyond the Return to Normalcy"; Voelker, "History of the International Brotherhood," 187–219, 368–78, and passim. See also Burke's addresses to delegates at Brotherhood conventions: e.g., Brotherhood, *Report of Proceedings . . . 18th Convention* (1939), 8–11.

had formed a benevolent society which sought the passage of laws that would reduce working hours in the paper mills. Eventually turning to trade union activities, the machine tenders barely weathered the discouraging 1890s and formed the nucleus of an organization chartered by the AFL in 1902 that was called the International Brotherhood of Paper Makers (IBPM).

The early organizations, however, represented only skilled machine tenders and paper workers. The less-skilled pulp and paper workers remained largely untouched by union organization until the summer of 1901, when activists in the pulp mills of New York State secured a federal charter from the AFL. Expanding rapidly in New York and New England, the pulp workers claimed more than 5,000 members in several federal unions by the following year. The AFL initially rejected their petition for separate chartering as a pulp workers' union and placed them in an expanded version of the IBPM, which then became known as the International Brotherhood of Paper Makers, Pulp, Sulphite, and Paper Mill Workers.

Personal animosities between leaders of the two wings of the new union, however, together with prejudice against the less-skilled pulp workers on the part of the paper makers, created constant turmoil. In 1905, paper makers voted to prohibit any representative of the pulp workers from become president of the organization, and the following year pulp workers' locals seceded from the enlarged organization. For the next three years, bitter conflict raged between the two organizations as each scabbed against the other, sought to undermine the rival union's influence among mill workers, and generally divided the potential membership of paper unions into factions. The rival presidents, Jeremiah Carey of the paper makers and James F. Fitzgerald of the pulp workers, engaged in intense personal attacks that exacerbated the conflict festering in the mill towns. By 1909, trade unionism in the industry — confined in any case largely to New York State and New England — had been exhausted as the promising beginning of the decade's early years gave way to scabbing, recrimination, and backbiting.[2]

Encouraged by the AFL, in 1909 the two organizations eventu-

2. Voelker, "History of the International Brotherhood," 28–49; Graham, *Paper Rebellion*, 1–5; "United Paperworkers International Union," and "International Brotherhood of Pulp, Sulphite, and Paper Mill Workers of the United States and Canada," in Gary Fink, editor-in-chief, *Labor Unions* (Westport: Greenwood, 1977), 277–80, 310–12.

ally signed a peace treaty. This document divided jurisdiction between them and established once and for all the separate iden-
tity of the Pulp, Sulphite, and Paper Mill Workers. Fitzgerald had
left the union in 1908, removing one of the main reasons for hostility. By virtue of the agreement signed in June, as confirmed by
the issuance of an AFL charter on July 2, 1909, the Pulp, Sulphite,
and Paper Mill Workers gained jurisdiction over the less-skilled
workers in the machine rooms and beater rooms, together with
"all other pulp and paper mill help who are not connected with
any other international union." Although the workers were
roughly divided between the two organizations according to skill,
in practice the Pulp, Sulphite, and Paper Mill Workers' jurisdiction embraced a wide spectrum – common labor in the woodsheds
and debarking operations, semiskilled grinders, deckers, and
screeners, and the relatively highly skilled cooks, acid makers, and
other workers in the digester room. Moreover, the precise termi-
nology of the union's jurisdiction in primary pulp and paper operations tended toward inclusiveness and expansion, benefiting the
Pulp, Sulphite, and Paper Mill Workers as technological and chemical changes changed work structures and job definitions. Thus,
while the union was limited to sweepers and wipers in the machine rooms, it could claim all employees except the engineers in
the beater rooms. "In all other departments . . . the jurisdiction
of this organization is paramount and inclusive," aside from skilled
craftsmen such as electricians, mechanics, and carpenters, who
belonged to their respective craft unions.[3] The jurisdiction of the
Paper Makers tended to remain stable, covering traditional, highly
skilled operations directly related to the running of the fourdrinier
machines, while the categories of jobs covered by the Pulp, Sulphite, and Paper Mill Workers tended to expand to accommodate
newer functions in pulp transport, preparation, paper handling,
and converting.[4]

3. Brotherhood, *Constitution and By-Laws* (n.p.: various dates; available on
microfilm in series entitled "Convention Proceedings and Constitutions of American Trade Unions," Reel 148).

4. Over the years, the Brotherhood gained jurisdiction over bag, box, envelope, and paper-handling workers (1922), insulite workers (1926), and virtually
every other category of converting plant worker (1935). Brotherhood leaders believed that these jurisdictions were implicit in the initial jurisdictional grant contained in the 1909 charter. See Brotherhood, *Constitution and By-laws*.

With the signing of the peace treaty, relations between the two organizations became more amicable. During the 1920s, however, differences in emphasis and character continued to divide the AFL paper unions. Pulp workers never entirely forgot their secondary status in the Paper Makers, and IBPM wage strategy, which stressed the goals of the skilled machine tenders and beater foremen in that union, antagonized the unskilled wood handlers, pulp workers, and deckermen who formed the heart of the rival brotherhood's strength. As long as Carey remained president of the Paper Makers, old animosities rankled, even after the accession of Burke to the presidency of the Pulp, Sulphite, and Paper Mill Workers in 1917. Still, through the late 1910s the two organizations remained in close contact, refrained from open hostilities, and even exchanged jurisdictions and bargained in common with many large pulp and newsprint firms.[5]

After World War I, however, both organizations fell on hard times. In the years following its chartering, the Pulp, Sulphite, and Paper Mill Workers had grown steadily, from a base of about 1,500 members in 1910 to more than 7,000 in 1917. Although an important strike against International Paper in 1910 had been instrumental in establishing the new organization's credibility, moderate union demands and the cautious leadership of John Malin, the union's president-secretary during these years, helped to maintain industrial peace. Although membership was overwhelmingly concentrated in New England as late as the onset of the war, the organization began cautiously to probe into the Great Lakes states, the South, the Pacific Northwest, and especially Canada, increasingly the center of newsprint production after the Canadian American tariff treaty of 1913 eliminated duties on Canadian newsprint entering the United States.

World War I brought boom times to the paper industry, and the union prospered. John P. Burke, who was elected president-secretary in 1917, continued Malin's cautious leadership, and by the early twenties membership had risen above 15,000. Wartime demand had stimulated new product development and tentative expansion of the industry into the South and along the West

5. Cernek, "Beyond the Return to Normalcy," 55–126; MacDonald, "Pulp and Paper," 103–12. Voelker, "History of the International Brotherhood," 106–108, stresses the accommodation between the two unions after 1909.

Coast. Since the turn of the century, International Paper, Great Northern, and other multipurpose concerns had been expanding and consolidating, moving into new product lines as newsprint production yielded to Canadian competition and as the war created new uses for paper products. Trade unionists in the paper industry, like their brethren in other sectors of the American economy, looked forward to postwar consolidation and expansion; the war years had been marked by cooperation between labor and management and by loyal and steady production on the part of the workers.

The postwar period, however, shattered these hopes. Employers in the paper industry began a concerted antiunion campaign coupled with a program of wage cutting. Apprehensive about the growth of unions under favorable government policies during the war, companies embraced the so-called American Plan, with its open-shop orientation. Some Wisconsin companies instituted versions of welfare capitalism. The Scott Paper Company consulted with famed labor economist John R. Commons to develop programs designed to instill employee loyalty through enlightened personnel policies. The Kimberly-Clark Company, a Wisconsin firm specializing in tissue products, established a widely publicized industrial council, joining the long list of firms in the 1920s that experimented with various forms of company unions and employee representation plans. Other concerns, such as Nekoosa-Edwards, on the Wisconsin River, combined a strong antiunion stand with sophisticated public relations campaigns designed to alert mill workers to the need for conservation and hygiene. Still others, such as Marathon Paper, under the leadership of D.C. Everest, one of the boldest and most imaginative of paper company executives in the postwar period, fought unions with no holds barred, recruiting labor spies and openly attacking the unions. Thus, if the paper unions expected the new era of the twenties to usher in a period of growth and acceptance, they found themselves grossly mistaken, for the idea that unions were disruptive, contentious, and inefficient elements in modern productive enterprise remained forceful in the paper industry.[6]

In addition to this pervasive antilabor atmosphere, the twenties spawned the greatest and most destructive of the strikes ever

6. Smith, *History of Papermaking*, 600–602.

to afflict the industry. "For several months in 1921," Burke later recalled, "we had two-thirds of our entire membership out on strike. Before these strikes were over we had lost the greater part of our organization and had all but exhausted our treasure."[7] In the spring of 1921, the unions clashed with giant International Paper over wages and work rules. In July of that year, the company opened its struck mills on a nonunion basis and vowed to impose the open shop, after having dealt with the unions since 1911 on a company-wide union shop basis. Then in 1922, the Pulp, Sulphite, and Paper Mill Workers struck a group of paper companies led by St. Regis, breaking with the Paper Makers and the skilled crafts over the companies' repeated efforts to slash wages.

The Pulp, Sulphite, and Paper Mill Workers had more than 15,000 members at the close of the war and fought vigorously. The twenties saw confrontation, injunctions, evictions, enormous membership and financial losses for the union, and drastically curtailed production and profits for International Paper and other companies. The fight was particularly bitter at International Paper's large, often antiquated New York mills. Early in the strike the two paper unions cut sharply into the company's production. In Palmer, New York, the site of a huge mill, striking workers met every week-day for five years to demonstrate their solidarity. "I doubt if any other body of organized workers anywhere in the United States can match this record," Burke proudly asserted.[8]

But in the end the company's power prevailed. Production gradually rose, and in 1922 the craft unions, representing strategically placed machinists, firemen, and electricians, went back to work, helping to break the back of the paper unions' struggle. The strike continued for five years, as International Paper seized the opportunity to close down obsolete mills and to impose important changes in work rules and payment schedules. For the unions, the strike ended in recrimination and antagonism that resounded for a long time indeed in the remote mill towns. The paper unions detested the behavior of the crafts. Even under the best of circumstances, paper union leaders regarded the Firemen, Machin-

7. Typescript of John P. Burke, Report to the 19th Convention, Sept. 8, 1941, Reel 9.

8. Cernek, "Beyond the Return to Normalcy," 127–214; Voelker, "History of the International Brotherhood," 187–229; Burke is quoted in Graham, *Paper Rebellion*, 6.

ists, Carpenters, and other craft organizations as parasites that claimed their respective clusters of skilled workers after the paper unions had done the tough organizing. Now, however, the paper unions accused the crafts of scabbing, raiding, and sellout as they recruited strikebreakers and disaffected pulp and paper workers.

In 1924 and 1925, Burke sought to bring the long strike to a close, hoping to preserve traditions of union security at International Paper by accepting significant wage cuts and changes in work rules. The company was obdurate, however, and striking workers repudiated Burke's efforts, accusing the international union of callous indifference and even sellout. For his part, Burke found that he could not raise strike funds from nonstriking locals and that the tendency of paper industry union members to refuse to look beyond the immediate confines of their particular place of employment seriously undermined the solid front that he and the Paper Makers sought to maintain.[9]

The great International Paper strike petered to an end in 1926. The union spent some $242,000 in strike support, an enormous sum for such an impecunious little organization in the 1920s. In the decade prior to the strike, the Brotherhood's union shop agreement with International Paper had covered twenty locals. By 1926, the organization no longer existed in the mills of the corporate giant. The union also suffered serious defeat in strikes against a group of paper makers led by St. Regis Paper Company in 1922; the membership took the brunt of sharp wage reductions. The organization declined from 15,000 dues-paying members in 1921 to some 6,000 in 1926 and was no longer a force with some of the largest and most dynamic firms in the industry.[10]

In the last years of the twenties, the union struggled painfully for sheer survival. Pulp and paper workers, while often evincing passionate solidarity, had never been eager dues payers. In the lean years of the late twenties, the organization cut its staff, reduced operations, and for a time even ceased publication of its journal. A companywide union shop agreement with the Great Northern Company and judicious expansion into newsprint plants

9. Cernek, "Beyond the Return to Normalcy," 151 and passim; Voelker, "History of the International Brotherhood," 187–219.

10. Cernek, "Beyond the Return to Normalcy," 210–26; Brotslaw, "Trade Unionism," 111–32; MacDonald, "Pulp and Paper," 112.

in Canada provided the thin margin of income that enabled the Pulp, Sulphite, and Paper Mill Workers, to remain afloat.[11]

The drastic business downturn of 1929–30, of course, struck with devastating force. "By the time we get through this spring," sighed Burke in 1930, "I fear that our organization will look like a ship that has been through some gale."[12] The organization's belt tightened. The budget was trimmed even further, and the few friendly employers who maintained union-shop arrangements were cherished. Never the most regular of dues payers, pulp and paper workers faced layoffs and short time; many stalwart fighters of earlier conflicts lapsed into apathy, although some were compelled by union security provisions to contribute to the union. As one of the union's small remaining corps of organizers reported in the dismal summer of 1930, workers in Wisconsin "say [that the] union is a good thing, but when we ask them to sign up and pay [it] seems . . . that . . . they hate to give a little money."[13] The mills were running part time, there was little money for organizing, and workers were demoralized and impoverished. Burke and his associates had long since abandoned the high hopes of the early twenties. "As I view it," the embattled president-secretary remarked grimly, "the only thing the unions can do during these times is to hang on and, above everything else, try to save our organization." The question haunted him: "What would happen to the workers if the labor movement should be destroyed?"[14]

By the coming of the New Deal, Burke's tenure as president-secretary had lasted sixteen years. He was born in Vermont in 1884 and entered the paper mills at an early age, becoming an active trade unionist by the time of his twenty-first birthday. In 1905, he joined the pulp and sulphite local of the International Brotherhood of Paper Makers at a mill in Franklin, New Hampshire. In 1906 he became a local officer, and he joined the international staff as a special organizer. By this time he had been recognized as a man of broad interests, judiciousness, and complete trade union loyalty, and he rose rapidly. In 1914 he became

11. Graham, *Paper Rebellion*, 9–10; Brotslaw, "Trade Unionism," 111–25; Voelker, "History of the International Brotherhood," 345–48; Sullivan to Burke, Dec. 22, 1933, Reel 1.
12. Burke to George C. Brooks, May 16, 1930, Reel 1.
13. R.C. Henke to Burke, June 22, 1930, Reel 1.
14. Burke to A.C. Krueger, July 25, 1930, Reel 1.

vice-president of the union and also of the New Hampshire Federation of Labor. In that year, he ran as Socialist party candidate for governor of the Granite State and in 1916 became the first vice-president of the Pulp, Sulphite, and Paper Mill Workers. When John Malin stepped down as president-secretary in 1917, Burke was unanimously chosen to be the new executive officer. For a period of forty-eight years his leadership was never seriously contested.

Throughout this time Burke commanded wide respect in the labor movement as a man of dedication, firm convictions, and integrity. Although the union underwent its share of controversy in both the dismal twenties and the ebullient thirties, he remained largely immune from criticism by dissidents. He largely eschewed the trappings of office that so many trade union officials found irresistible. He regularly returned a small percentage of his modest salary to the union coffers and had no taste for high living. He neither drank nor smoked, and he and his small office staff conducted the union's affairs from a small suite of offices above a store in Fort Edward, a mill village with a population of about 2,000. "I am," he declared in 1939, "a little known individual residing in an obscure village in Northern New York."[15]

By 1941, when the union had more than 60,000 members, the office force at Fort Edward consisted of seven or eight secretaries, messengers, and accountants. John Burke's wife, Bessie Leon Burke, served as his paid assistant and handled much of his voluminous correspondence. In some years he traveled as much as 30,000 miles, most of it by rail, although by the mid-thirties he was making an annual multihop flight to the Pacific coast to meet with the growing locals in that important region. Burke was a literate man who frequently sent books on labor history and public affairs to his many correspondents in the labor movement and on the manage-

---

15. "John P. Burke," in Gary Fink, editor-in-chief, *Biographical Directory of American Labor Leaders* (Westport: Greenwood, 1974), 45; *Pulp and Paper Worker* 22 (Jan. 1965) (special ed.); Matthew J. Burns, "My Friend—John Burke," *Pulp and Paper Worker* 24 (May 1966); Voelker, "History of the International Brotherhood," 357–78; Cernek, "Beyond the Return to Normalcy," 58, 94–96, 223–26, and passim. Monthly Financial Reports, 1941 (e.g.), Brotherhood Papers, 1941, Reel 1, suggest the scope and style of Burke's leadership activities. See also Burke to William Hard, Nov. 29, and Press Release, June 28, 1941, both Reel 9, and Burke to J.B. Rooney, Sept. 1, 1940, Reel 4, regarding his personal habits. The quotation appears in Burke to Morris L. Cooke, Nov. 22, 1939, Reel 1.

ment side. He read the *New Leader* and other social democratic publications, as well as the trade union press, although he complained that his manifold duties left him little time for keeping abreast of current events. "When I worked in the mill ten and twelve hours a day I used to have time to do a little reading and be half way informed,"[16] he once remarked. He handled much of the union's enormous correspondence personally, particularly relishing opportunities to discuss broad social issues with certain employers, socialist intellectuals, and old friends in the labor movement. He also enjoyed attending an occasional baseball game (he had been, his colleague Matthew Burns of the Paper Makers recalled, a pretty fair amateur ball player in his youth), visiting his mother in New Hampshire from time to time, and, above all else, spending time in the garden at home in Fort Edward. But for the most part, Burke's life was his work. "I have been on this job for twenty-one years," he observed in 1938, "and during all that time I have been on duty every day and every night." He did spend one day bedridden following a bout of food poisoning.[17]

Through the years of his leadership, Burke espoused a social philosophy compounded of trade union pragmatism, humanitarian socialism and equally nonsectarian Christianity. His commitment to trade unionism transcended narrow considerations of specific advantages to be gained by virtue of union membership. To him, the union was an end in itself, a singular contribution of Western, democratic society. The free participation of workers in organizations of their own choosing, and the financing of these organizations by the workers themselves, not only ensured the workers protection in the economic realm but provided society with a vigilant outpost of free men and women, beholden to no government or church or party.

Burke believed that organizers too often stressed only the prospect of immediate financial gains in appealing to new recruits. Such a policy was dangerous, for not only was it unlikely that the union could unfailingly meet the workers' heightened expectations, but workers might lose sight of the fact that the primary consideration must always be the maintenance of the union itself, even if its preservation occasionally required strategic retreats in regard

16. Burke to William Bohn, May 20, 1938, Reel 7.
17. Burke to Robert B. Wolf, April 11, 1938, Reel 7.

to wages or conditions of employment. In the years since the head-strong policy of taking "no backward step" in the 1920s, Burke told the 1941 convention, "we have tried to pursue a different policy. When unions are led by skilled and experienced men, . . . retreats are often . . . good strategy."[18] In the same vein he had earlier editorialized that "we must not stress too much what they are going to get from membership in a labor union. . . . The workingman who pays an initiation fee . . . with the sole thought . . . of getting a raise in wages next week," he declared, would never learn the fundamental lessons of solidarity, loyalty, and social idealism that were the true benefits of the labor movement.[19]

Unfortunately for this vision, pulp and paper workers did not always conform to Burke's high-minded strictures. Support for the paper unions had historically been sporadic and inconsistent, flaring into obstinate militancy on some occasions and lapsing into apathy and frustration on others. Without the formal apprenticing and high initiation fees of the craft unions, the paper unions, even among relatively skilled mill workers, had to organize on the job and never enjoyed the stability of membership that the closed shop offered the crafts. Union organization and labor relations in the mill towns tended to be highly personal, with both company managers and union leaders subject to microscopic scrutiny on the part of the workers. A dynamic and benevolent employer could undercut the unions' appeal by means of careful public relations and colorful gestures, especially in view of labor's grim and so frequently unsuccessful struggles throughout the first third of the century.

On the other hand, pulp and paper workers evinced some of the solidaristic élan often characteristic of workers in physically isolated locales. Support for the union was perhaps less than a reflex, but these workers had certainly waged their epic struggles. The five-year fight against International Paper in the 1920s was merely the latest in a series of strikes, often noteworthy for the workers' stamina, dating to the eighties. The willingness of workers to support a strike, Burke observed, "presents a trade union phenomenon that I have observed time and time again" in the

18. Typescript copy of Burke, Report to the 19th Convention, Sept. 8, 1941, Reel 9.
19. Burke editorial, *Brotherhood Journal* 18 (Jan. 1934), 1.

pulp and paper industry. "When there is a battle of any kind, the majority of the workers made good union men. They are fighters."[20] Burke was particularly proud of his members' gallant, protracted battle against International Paper in the twenties, despite the follies and defeats that eventually attended it.

But alas, the life of the union member was not a series of stirring battles. Parades and picket lines eventually gave way to membership meetings, dues collecting, and routine union business. These activities inspired Burke's fear for the labor movement and for the future of his own cherished organization. "When the battle is over and the excitement has died down, then a relapse takes place, and the majority take a lackadaisical interest in the union," he complained.[21] He exhorted workers to attend meetings, to pay their dues, and to become informed about the day-to-day operations of their local unions. Workers too often found union business dull and unexciting. "They are all out duck shooting in Green Bay," he lamented to a distressed organizer one day in 1930. "It seems that the workers always have something to take up their attention."[22]

Whether attempting to win new recruits or coaxing members to pay dues and attend meetings, Burke and other union representatives found pulp and paper workers fickle and unresponsive. Indeed, it sometimes seemed to the veteran president that the behavior of workers in the optimistic days of the thirties compared unfavorably with the sacrifices that union men had endured in less propitious times. "All workers in those days joined voluntarily," he informed a local unionist in 1941, "they joined oftentimes in face of threats. . . . Can it be possible that American working men have slipped backward so much in the last twenty-five or thirty years that we now cannot have a union unless it is based upon compulsion? Frankly I hate to think that this is true."[23]

It galled Burke that his union, in both good and bad times, depended on the union shop for a stable membership and for regular dues payments. It was only through such arrangements with

20. Burke to John F. Wheeler, June 29, 1939, Reel 4.
21. Ibid.
22. Burke to George C. Brooks, Oct. 31, 1930, Reel 1.
23. Burke to Milton Bever, Dec. 1, 1941, Reel 5.

Canadian paper companies and a few American concerns in the late twenties and early thirties that the organization had been able to remain in existence. The union worker, Burke held, should pay his per capita tax cheerfully, for it tied him to the great traditions of the labor movement and freed him from dependence on political parties, churches, or governments. Yet to his consternation, workers did not welcome the thought of parting with the two-dollar initiation fee or the per capita tax each month. "I think one of the things that will puzzle the labor historian of the future," he wrote in 1941, ". . . will be the fact that working people needed compulsion to make them loyal to the union."[24] When local unionists complained about no-strike clauses in contracts and other concessions to employers, the president-secretary replied, "If the workers would only organize without [union shop] agreements and remain organized then we would not have to be bothered about living up to agreements. However, just as long as American workers have to have the employers keep them in the union," they will have to accept disagreeable limitations on their freedom of action.[25] It was necessary for the president-secretary or his organizers to enforce unpopular contract clauses despite local dissidence in order to maintain union shop provisions. "It seems," Burke consoled an organizer faced with difficult problems of contract enforcement, "that we have to be dictators sometimes or the workers will destroy themselves."[26]

To Burke, the whole panoply of trade union practices constituted the bargain that workers had to strike for an independent labor movement. Without embarrassment, he affirmed the collective-bargaining contract, no-strike provisions, and the union's duty to discipline members. Just as the union had an obligation to press grievances against chiseling employers, so it had a duty to ensure that workers gave a fair day's work in return for their pay. Burke felt a certain admiration for the Industrial Workers of the World (iww), who rejected the labor contract and advocated

24. Initiates normally paid two dollars, of which half went to the local treasury and the rest to international headquarters, where it paid for their first month's dues. Monthly dues to be paid to the international union were 60 cents for men, 40 cents for women. The quotation appears in Burke to Walter Harder, July 15, 1941, Reel 5.
25. Burke to Morris Wray, July 24, 1939, Reel 3.
26. Burke to S.A. Stephens, Jan. 30, 1941, Reel 1.

direct job action, but he argued that unless bona fide unions were willing to accept the outcast status that the IWW had willingly embraced, they would have to abide by, and to enforce, the contracts they signed.[27] Through all of the years of struggle and eventual success, in his eyes the union, itself of paramount concern, remained in bad times and good the only vehicle through which the worker could seek justice in the workplace and an independent and uncompromised voice in industrial society.

This intense trade union commitment moderated and relegated to the background Burke's socialist views. He deplored the European practice of uniting the unions with ideological political parties; he regarded the ideological factionalism of the labor movement in Germany as a key factor in Hitler's rise to power. He had run as a Socialist for governor of New Hampshire in 1914 and kept in touch with trade union socialists and socialist intellectuals associated with the Rand School in New York. He also read social democratic publications and corresponded regularly with old comrades. The Democratic party of Franklin D. Roosevelt did not arouse his enthusiasm; he shared (ironically, in light of his distaste for the Mine Workers' chieftain) John L. Lewis's views that labor would be endangered if it became part of a partisan political apparatus. He regarded FDR as well intentioned but dangerous, too powerful for a republican citizenry: government policies might aid the worker only momentarily, and the labor movement could easily be betrayed or assailed, leaving workers without independent footing. Government spending elicited his skepticism; in his view, encouragement of the private sector was the key to reducing unemployment.[28]

At the same time, his critique came from the Left, not from an intrinsically probusiness viewpoint. He had contempt for both Republicans and Democrats; on one occasion he said, "I never voted those tickets." In 1924, he had supported Robert M. La Follette, and in the thirties he cast his ballot for Norman Thomas. "I was personally acquainted with Debs," he recalled proudly. "I spoke from the same platform with him many times. I knew all

27. Burke to Executive Board, Nov. 5, 1937, Reel 3.
28. Burke to Robert Wolf, Nov. 8, 1936, Reel 4; Burke to Morris Wray, Dec. 7, 1939, Reel 3; Burke to John Bayha, Aug. 28, 1940, Reel 4; Burke to E.W. Kiefer, Feb. 4, 1941, Reel 10.

of the [six] Socialist assemblymen who were unseated in the [New York State] assembly" in 1920. "Nearly all of the old time socialist leaders were imbued with lofty ideals."[29] He revered Debs, finding him a truly Christ-like man who embodied the best in social idealism, selflessness, and personal magnetism.

He believed, however, that present-day socialists were unrealistic in their expectations of the labor movement and failed to understand the practical realities under which it toiled. His socialist comrades, he complained, constantly pilloried the labor movement for failure to organize the unorganized, but they failed to realize that the task was arduous, requiring careful attention to detail and involving as much cajolery as militant posturing. "I would like to have some of these Comrades have my job," he told William Bohn, secretary of the Rand School, in 1938. Not only did he have to direct the renewal of more than 200 agreements in a period of recession and wage cutting; he also had to deal with workers of all nationalities scattered from British Columbia to Florida, "large numbers of whom cannot speak a word of English." How would his friends in the socialist movement deal with a new local of Japanese-speaking workers in Ocean Falls, B.C.? he asked. He had made them happy by having the union constitution printed in Japanese, "but, will they give me their support after our convention . . . adopts a resolution of sympathy with the Chinese? Woe and more woe!" Socialists accused the labor movement of discriminating against Negro workers, but the local unionists in the South — aggressive members of the rank and file, normally the darlings of the Left — constituted "the real source of opposition to admitting negroes [*sic*] to membership," Burke informed Bohn.[30]

Burke was, of course, anti-Communist. He expressed his antagonism toward the Communist party in countless communications and many editorials in the *Journal*. He urged correspondents to read *Out of the Night*, the exposé of Stalinist terror.[31] At the same time, through the 1930s the union did not constitutionally bar membership on the basis of political party or ideology. Local 107, a converted paper group in New York City that had been affiliated with the Brotherhood since 1917, had a leftist leadership in

29. Burke to Carl Caspar, Jan. 4 and June 13, 1941, Reel 9.
30. Burke to William Bohn, Aug. 8, Nov. 7, 1938, Reel 7.
31. "Jan Valtin" (Richard Krebs), *Out of the Night* (New York: Alliance, 1941).

the 1930s, but otherwise there was little organized Communist influence in the union. Burke, in his capacity as editor of the Brotherhood's *Journal*, made no effort to censor communications from this local, nor did he make overt attempts to stifle organizers who evinced sympathy with Communist party positions. Throughout the mid to late 1930s, he carried on an extensive correspondence with this small group of leftist organizers and local union activists, constantly commenting on the sins of the Soviet Union, the destructiveness of Communist elements in the new Congress of Industrial Organizations, and the virtues of open, democratic societies. At the same time, he appeared not to regard them as a personal or institutional threat and remained content with counseling them to moderation while he scored debater's points as the American Communist party thrashed about in apparent response to Soviet directives.[32]

Burke no doubt regarded these occasional debates with leftists as entertaining, a counterpoint to the knotty problems of building and maintaining a growing union in the thirties. The grandiose hopes of the Left for final victory over capitalism and the radicals' hope that American workers would rise to the barricades seemed ludicrous in light of the practical problems he encountered in his daily contact with pulp and paper workers and their locals. In particular, Burke felt that he had learned some important lessons in the desperate days of the 1920s, lessons that both modified and reaffirmed his basic trade union commitments.

The strikes and economic disasters of the twenties and thirties reinforced Burke's determination to maintain the union at all costs. In addition, they persuaded him even more firmly of the need for caution and restraint. He had seen much folly in the union's behavior in the strikes of the past decade. "We spent thousands of dollars on lawyers during the strike . . . because of a great many foolish acts on the part of some of our members," he later recalled.[33] Discretion sometimes seemed the better part of laborite valor. "Instead of adopting a realistic attitude and deciding on an orderly retreat," he mused on another occasion, ". . . we decided on a pol-

32. Burke's correspondence with Hyman Gordon, president of Local 107, 1933–41, which is found in the yearly file for Local 107, and with Morris Wray, 1937–1941, found in the yearly file for Wray, contains much of this material.
33. Burke to S.A. Stephens, Jan. 19, 1938, Reel 1.

icy of 'no backward step,' with the result that we ... took ... a great many of them."[34] Addressing the union's convention in the midst of successful organizing in 1941, he lectured the delegates. "We have learned that in bad times it is sometimes better for the army of labor to execute an orderly retreat rather than have our union ranks decimated by strikes when economic conditions make it almost impossible to win."[35]

In the heady days of the thirties, militants sometimes chided him and his veteran associates for failing to employ mass organizing techniques or to enlist under the CIO banner. But the president-secretary observed to his more exuberant members that the organization had experienced sudden spurts in membership before only to see its numbers melt to a core of stalwarts. In 1937, for example, he noted that union officials had been encouraged by recent membership growth but that the union had had a larger membership seventeen years earlier, before the disastrous strikes of the twenties. "If the International Officers appear, at times, to be too careful, too conservative, too cautious," he advised the ranks, "remember that years of experience have taught them" of the dangers of haste and precipitate actions.[36]

The defeats of the 1920s, however, had left positive lessons as well. They persuaded Burke ever more firmly of the absolute need for industrial and semiindustrial forms of organization. The behavior of the crafts during and after the strikes, and the unwillingness of the powerful AFL unions to organize mass production workers, caused him in the early New Deal years to side consistently with Lewis, David Dubinsky, and the other industrial union supporters in AFL debates. Burke's tiny Brotherhood had constantly been victimized by the crafts, which could use their enormous weight in the AFL to secure favorable jurisdictional rulings. In a typical rueful comment, Burke remarked to his Wisconsin organizer in 1930, "I presume that after we spend a great deal of time and money at Stevens Point and get a union shop agreement, that the Machinists will then come along a disrupt our local."[37] Clearly,

34. Burke to George Roller, Sept. 29, 1933, Reel 3.

35. Typescript of Burke, Report to the 19th Convention, Sept. 8, 1941, Reel 9.

36. Burke, Editorial, Brotherhood *Journal* 20 (April 1936), 3.

37. Burke to R.C. Henke, June 5, 1930, Reel 1.

he remarked, "it requires a union like the Pulp, Sulphite and Paper Mill Workers to really organize the thousands of low-paid workers in this industry."[38] By the late 1930s, after the CIO challenge and the mass organizing campaigns in steel, rubber, autos, and other mass production industries, Burke reiterated the counsel that he had given AFL chieftains all along: "One thing is sure – and the other officials of the American Federation of Labor must realize it – there must be greater opportunity permitted . . . for the industrial or the semi-industrial form of organization."[39]

The other major lesson was that the union had to cooperate with industry. There was no point in harboring resentment over the affairs of the twenties. Bitterness and recrimination would not organize International Paper or the other major concerns. At the same time, smaller companies had loyally kept union shop agreements when they might have sought to eliminate the union. Certainly, it would not now behoove a triumphant organization to press unreasonable demands upon these firms. Even during a phase of union growth, it was important for workers and officers to demonstrate to employers the benefit of having a responsible labor organization. "Many employers today encourage their workers to join our unions," proudly noted Jacob Stephan, one of the union's most trusted representatives. "Let us show the company that it pays to deal with our organization."[40]

Burke made every effort to stress his own and his union's understanding of industry's problems and the contributions of the paper companies to the workers' well-being. "I believe that workers must be taught that their welfare is inseparably bound up with the welfare of the company," he reassured one executive.[41] To another he declared, "Without the men of genius who made this industry possible, there would be no pulp and paper mill unions. Therefore, why not once in a while express a word of appreciation of the efforts of these forgotten men of industry?"[42]

Collective bargaining was, to be sure, often difficult, but in general, Burke's approach stressed reason, mutual advantage, and

38. Burke to Fred G. Hoodwin, Dec. 22, 1939, Reel 7.
39. Burke to Cong. George Schneider, May 1, 1938, Reel 7.
40. Brotherhood *Journal* 22 (July-Aug. 1938), 13–14.
41. Burke to H.I. Houlette, Vice-President, Ohio Boxboard Company, March 4, 1938, Reel 7.
42. Burke to E.D. Stoetzel, Marathon Paper Mills, Jan. 13, 1941, Reel 10.

openness. "We still find among the business and professional class many who would . . . deny the workers the right to organize,"[43] he noted, but he nonetheless found the managers and owners of the larger mills well informed and willing to deal responsibly with responsible leaders. The great International Paper strike of 1921–26, however productive of heroic legends and mill-town mythology, had been a dreary bloodletting, for the company as well as for the union. Confrontation and class conflict had had their inning in the paper industry; now it was time for accommodation and mutual understanding. "The only kind of a labor agreement that is mutually beneficial," he advised one of his lieutenants, "is when both parties are smiling and cracking jokes when they are attaching their signatures to the agreement."[44]

More militant elements in the industry sometimes criticized Burke for being too accommodating. Still, he could not be fairly accused of desiring industrial peace at any price. Like Lewis and other leaders of industrial unions, he felt that strikes in large, mass production segments of the economy were an entirely different matter from the controlled, carefully targeted work stoppages characteristic of construction sites. It was sometimes possible to play off one operator against another in decentralized bargaining, but when wage cutting began, the union might have to adopt a defensive stance and try to prevent employers from whipsawing the union even as it accepted temporary downward adjustments. The history of the industry before 1933 convinced Burke that the union was, for all of its growing strength in the 1930s, still a delicate organism whose nurture required care, patience, and even occasional pruning.

Nor did Burke believe, in the final analysis, that collective bargaining per se could solve the "labor problem." Despite many years of mutual accord with a number of concerns, he remarked in 1941, workers "still think of the company . . . in the light of an oppressor." Collective bargaining and labor contracts were not "some kind of religion." In fact, he admitted to disliking the very term "collective bargaining." The Clayton Act of 1914 seemed to have declared that labor was no longer to be considered a commodity, but in the next round of negotiations, conducted no doubt

43. Burke to Helen Safferstone, March 16, 1933, Reel 3.
44. Burke to John Sherman, April 26, 1941, Reel 1.

in good faith, "we shall bargain to sell the collective labor power of our . . . members at the highest possible price." However necessary these negotiations proved to be in the imperfect world, and however great an improvement they represented over the arbitrary power of employers, Burke still found the process unfair and degrading to the cause of human dignity.[45]

Burke's critique of collective bargaining never reflected a full-fledged ideological perspective. His socialism remained a general, even vague, non-Marxist humanism that stressed the need for sacrifice and equity. He sometimes identified the "acquisitive instinct" as the source of labor's and society's woes. He challenged one employer: "Now suppose that employers . . . would rid themselves of this acquisitive instinct, this desire to pile up millions and millions of dollars." He challenged employers to emulate Jesus Christ, as he felt Eugene V. Debs had done. He looked to a time when Christian principles of common humanity would replace avarice. "The dividing line between management and union labor, under such a setup, would be eliminated. Ownership and management and workmen would then be one team, pulling together . . . , sharing."[46]

But John Burke rarely indulged in his visionary dreams. Normally he was quite content to cope with daily life, to adjust to practical realities. A complex and sensitive man, he had a strong awareness of history and of the limitations under which he and his organization worked. In the world as it was, unions recruited members and retained them, often through theoretically objectionable union shop arrangements. Unions made and honored bargains despite the occasional resistance of some members. They bargained sensibly, cherishing small gains, and hoping to avoid major losses. They recognized that their health and the welfare of their members could not be separated from the health and welfare of the industry. Burke saw unions as one of the prime engines of modern society for the advancement of the species. Unions were also one of the grandest manifestations of the dignity of the human spirit. These high-minded sentiments, however, did not prevent him from striking the bargains that needed to be struck

45. Burke to Robert B. Wolf, March 20, 1941, Reel 10.
46. Ibid.

and from acknowledging the limitations under which workers and their unions operated.

With the coming of the New Deal and the passage of the National Industrial Recovery Act and its Section 7(a), Burke and his cohorts stood ready to expand their organization. They were industrial unionists and were free of the hauteur that crippled many of their colleagues in the AFL in their encounters with mass production employees. Despite many setbacks, Brotherhood officials believed in labor's mission to organize the unorganized, and they found themselves armed with an uncommonly broad jurisdiction that gave them warrant to do so. At the same time, previous defeat and familiarity with dire circumstances instilled in them caution and restraint. With this ambiguous legacy the Brotherhood faced the challenges and opportunities of the 1930s.

4

Using the National Industrial Recovery Act in the Paper Industry

THE PASSAGE OF THE NATIONAL INDUSTRIAL RECOVERY ACT IN 1933 triggered intense union activity throughout the pulp, paper, and converted paper industry. The Pulp, Sulphite, and Paper Mill Workers had been virtually moribund prior to the appearance of Section 7(a), but they now found a steady source of recruits in older locals, in newer production areas in the South and the Pacific Northwest and, most surprisingly, in the hitherto dormant converted paper field.

President Burke and his lieutenants welcomed the new-found opportunities. The union enlarged its slender staff of organizers, making special efforts to resuscitate older locals. It worked with local activists, AFL organizers, and city central representatives to capitalize on the opportunities now available in converted paper. Burke and Vice-President Herbert W. Sullivan participated in National Recovery Administration (NRA) code hearings and served on labor advisory boards. Workers in the industry were enthusiastic, and the AFL affiliates responded. Membership grew. The union achieved footholds in geographical areas and in product sectors that it had never before entered. At the same time, the important NRA period revealed significant limitations of the union's organizing techniques and strengthened its leaders' convictions regarding the difficulty of welding the labor force in such varied industries into stable, effective unions. Both critics of the union's leadership and the union establishment emerged from the NRA period with regret that the opportunities to organize had been incompletely realized.

The upsurge of unionism in the early summer of 1933 startled Brotherhood veterans. Prior to the passage of the new legislation, the union's office had been virtually inactive. Twelve years of defeat and depression had caused retrenchment and retreat. Total union membership in the paper industry, embracing the IBPM and the crafts as well as the Pulp, Sulphite, and Paper Mill Workers, stood at no more than 10,000 in the United States and Canada combined. Most of those who had been organized worked in the newsprint segment, the least visible of all the components of the pulp and paper industry in the United States. John Burke found the lack of organization perplexing, for, as he observed in February, clearly "everything that the Socialists predicted in the way of breakdown of the capitalist system is coming to pass." Still, "the great mass of wage earners seem[s] to be as difficult to reach as ever." The most optimistic estimates placed Brotherhood strength at 4,000 on the eve of the New Deal.[1]

Yet even before the signing of the National Industrial Recovery Act on June 16, Burke stood poised to take advantage of the unusual opportunity. As an AFL affiliate with traditions of industrial unionism, the Brotherhood joined such organizations as the International Ladies' Garment Workers, the United Mine Workers (UMW), and the Amalgamated Clothing Workers in welcoming the opportunity to expand. At the same time, the union's peculiar circumstances ensured that President Burke would not attempt to follow the blitzkrieg methods that John L. Lewis had employed in the coalfields. Made cautious by temperament and by previous defeat, Burke hesitated to commit the organization's meager resources in the hope of massive response. Skeptical of the pulp and paper workers' "gumption to organize," Burke added to his staff slowly. He relied on local veterans to rekindle the spirit in the older mill towns, scarred as they were by memories of de-

1. The membership estimate appears in Burke to Gustav Hitler, Secretary of Paper Making Workers' Branch, Verband der Fabrikarbeiter Deutschlands, Feb. 27, 1933, Reel 3. The words quoted appear in Burke to Mahlon Barnes, Feb. 22, 1933, Reel 3.

feat and betrayal, and employed his few staff organizers very selectively. Still, union officials clearly considered the government
initiative a once-in-a-lifetime chance to rescue their foundering
organization. "We should try to organize everywhere possible now,"
declared international auditor George Brooks, "because the workers have the protection of the Recovery Act."[2]

Response to the new opportunity to organize came from every
section of the country. In New England and New York State, old
union men cautiously began to speak out. In Wisconsin, long an
open-shop bastion, organizer Jacob Stephan probed into the heart
of antiunion territory, seeking to build local unions in the Marathon Mills of Wausau, the Nekoosa-Edwards mills further south
on the Wisconsin River, and the Kimberly-Clark plants, which
had well-established company unions, in Kaukauna and Menasha.
In Ohio, paper box workers, who had not previously been attached
to the Brotherhood but who fell under its jurisdiction, sought out
President Burke. Inspired by the activity of nearby rubber and
auto workers, the Ohioans asked local AFL officials what union
they belonged to and began to sign up their shops and to prosely-
tize neighboring workers before even asking the international
union about the procedures they should follow in seeking a charter. "I will never be satisfied until I see every man and girl working in a Box and Paper factory in this district signed in our union,"
vowed one such local activist.[3]

Even the remote Pacific Northwest and the alien South proved
responsive. In Washington and Oregon, the union spirit pervaded
the pulp and paper mills, with little direct encouragement on the
part of the international office. "We are a bunch of real fellows
here, not radical, but with an earnest desire to really improve our-
selves and working conditions within reason," wrote the secretary
of new Local 153 in Longview, Washington. From Alabama, North
Carolina, Louisiana, and elsewhere in the South came requests
for organizers, reports of early progress, and pledges of steadfastness. "We are getting serious about organized labor," the secretary

2. Brotherhood *Journal* 17 (June 1933), 3; Burke to George Brooks, June 10,
1933; Burke to Bart Doody, Oct. 17, 1933, and Brooks to Burke, July 8, 1933, all
Reel 1. See also Graham, *Paper Rebellion*, 14–18; Brotslaw, "Trade Unionism," 133–
34; and Smith, *History of Papermaking*, 445–51.
3. Edward Mangan to Burke, July 30, 1933, Reel 1.

of a new local in Louisiana declared, "and we want to go about it in a businesslike manner."[4]

The international union tried to respond to this unaccustomed demand for its services. "I have so many calls for organizers," Burke declared in July, "that I have neither the men nor the money to take care of all of them." By August the union was running out of dues books, and the printer could not keep abreast of orders for per capita stamps.[5] Well into 1934, organizing enthusiasm continued to tax the union's resources. In January, special organizer Rasmus Anderson, excited and frustrated at the same time by organizing gains in Wisconsin and by the union's inability to respond to all requests for help, bubbled over with impatience. "Oh," he ejaculated, "there are any number of places ready for organization around here!" Good news from the West Coast continued, and converted and paper box locals in the Middle West and along the East Coast spread organization into previously unaffected locales. "The entire valley is alive with unionism and every one is working harder than ever," wrote the secretary of NIRA Local 159, newly established in Lockland, Ohio.[6] By the end of 1933, the union had grown from an average monthly membership of 5,700 in 1933 to 13,500 in 1934. Burke doubled his organizing staff and employed on a temporary basis a dozen or so local unionists for special work. There were fifty-three local unions at the end of 1932; in 1933 the international added forty-five and in 1934 another thirty-two.[7]

Workers surged into the Brotherhood in 1933 and 1934 primarily in hopes of increasing their wages. All sectors of the industry had stood below national industrial averages for employee compensation even during good times, and the depression had taken its toll from pay envelopes. Pulp and paper companies, while largely

---

4. George Clausen to Burke, Sept. 17, 1933; S.J. Roundtree to Burke, Dec. 15, 1933, Reel 2.

5. Burke to W.H. Wilson, July 24, 1933; Burke to H.W. Sullivan, Aug. 8, 1933, both Reel 1.

6. Rasmus Anderson to Burke, Jan. 10, 1934, Reel 1; Ella Mae Griffin to Burke, April 14, 1934, Reel 3.

7. Membership figures are in *Report of Proceedings . . . 16th Convention* (1935), 25; the numbers of locals and organizers are drawn from the index to the Brotherhood Papers, Labor-Management Documentation Center, Catherwood Library, New York State School for Industrial and Labor Relations, Cornell Univ.

avoiding massive layoffs, had reduced hours and had instituted work-sharing programs, all of which sharply reduced take-home pay for workers. Monthly compensation in 1933 for employed workers was $80, as compared with $117 for 1929, when about 20 percent more workers found employment in the mills.

In converted paper, workers complained about the faster pace of work and about diminished income. In the more competitive branches of the industry, employers sought to compensate for dwindling demand by increasing production and reducing labor costs in desperate hopes of gaining even momentary competitive advantages. In Ohio, male paper bag workers received even less in 1933 than had female operatives in 1929, despite the 20 percent sex differential in wage rates. Common labor in most pulp, paper, and converted paper facilities earned rates of less than forty cents an hour, yielding weekly incomes of less than sixteen dollars, barely half that which the Bureau of Labor Statistics deemed necessary to support a family of four. In the South and in backwater enclaves of several large cities, some women worked fifty-six-hour weeks and received as little as sixteen cents per hour in compensation. Throughout the converted paper industry, workers complained of chiseling as employers arbitrarily changed rates, downgraded skilled employees without changing their tasks, and substituted women for men. The entire industry seethed with discontent "from one end of this country to the other" over the question of wages, Burke warned NRA authorities in September 1933.[8]

Discipline on the job, deterioration of working conditions, and callousness on the part of managers also impelled pulp and paper workers toward the union. So overwhelming were economic concerns, however, that these recurrent problems received relatively little attention on the part of workers in the early New Deal period. Set-up paper box workers criticized box manufacturers for the quickened pace of work and for the dirty and unsafe character of the shops. Surely the president of a paper mill local in New

8. On wages, see Maclauren, "Wages and Profits," 199, 205, 207–208; Riley, "Earnings and Hours," 30; [Baril and Jones], comps., "Wages . . . in the Set-Up Paper-Box Industry, 1933 . . . 1935"; Jones, "Paper Industry Study," 213–14; Charles Vonie, Statement, Hearing 406-03 (Set Up Paper Box Industry), Oct. 3, 1933, NRA Records, Records Maintained by the Library Unit, Transcripts of Hearings; John P. Burke, Statement, Hearing 401-1-01 (Paper Bag Manufacturing Industry), Nov. 2, 1933, ibid.

York was not alone in seeing union rebirth as a means of ending "the petty tyrannies of our supts. and foremen during the depression." Shop-level issues, however, remained distinctly in the background as the pulp and paper workers articulated the reasons for their enthusiasm.[9]

The broader goals that workers brought to local union formation usually had less to do with commitment to organized labor than with a desire to support the New Deal, President Roosevelt, and the NRA. At least two new locals chose names reflecting this commitment, Local 190, representing paper box workers in the Boston area, which took the sobriquet "New Deal," and NIRA Local 159 in Ohio. No corps of organizers rolled through the mill towns and box shops claiming that "the President wants you to join a union," but throughout the industry workers identified enthusiastically with Roosevelt's program. "We have a man at Washington," exhorted a Virginia unionist, "the greates[t] man who ever entered the portals of the white house who is saying to you go to it boys organize 100%. . . . he is no other than Franklin D. Roosevelt." It remained for workers to "help him to make this recovery act a success." People in Ohio recalled that when the country had been in its greatest peril, Roosevelt had appeared on the scene, urging the people to work together for recovery. "Then came NRA—'We Do Our Part,'" reminisced one of Local 159's founders. "And realizing we could do our part most effectively by concerted action this local was formed." Workers in every sector seized upon the NRA promise, as a means both of bettering their individual lot and of revitalizing the country. Patriotism fused with self-interest. Some employers in Maine labored long and with only partial success to convince their skeptical Yankee and French Canadian employees that the NIRA did not *require* union membership and that the workers could, if they chose, fulfill their responsibilities to the recovery program as individuals or through employer-sponsored plant associations.[10]

The actual process by which new locals were organized and older organizations were revived varied in these heady days. In the mills

9. Vonie, Statement, Oct. 3, 1933, NRA Records; Tom Houston, President, Local 82, Tonawanda, N.Y., to Burke, Oct. 17, 1933, Reel 3.

10. Edward F. Eggleston, Local 152, Covington, Va., to McClellen Biggs, Dec. 5, 1933, Reel 1; W.E. Bachman to Brotherhood *Journal*, Sept. 15, 1934, Reel 3; Sullivan to Burke, July 20, 1933, Reel 1.

in New England and New York, Burke relied on veteran local unionists, dispatching Vice-President Sullivan and international auditor George Brooks to key locals at critical points in the organizing process. In Wisconsin, organizer Jacob Stephan sought to crack open-shop strongholds with mass meetings, visits to workers' homes, and constant proselytization in the mill communities in the Fox and Wisconsin river valleys. Through much of the converted paper industry and in both the South and the Pacific Northwest, organization began among the rank and file, as workers formed local groups, seeking out union representatives only afterward. These new union areas spawned a new group of union activists. Some had previous experience in other unions, but many were completely unfamiliar with organized labor, although they were eager to fight for social justice and, often, for personal advancement.[11]

In some pulp and paper mills, skilled craftsmen took the lead. Millwrights, firemen, electricians, and machinists even in non-union mills had often had experience with organized labor and grasped the opportunity presented by Section 7(a) to preach the union cause. In other cases, union workers in the nearby railroads, machine shops, or printing industry (in urban parts of the converted paper industry) took it upon themselves to spread the union appeal and served as initial counselors for inexperienced pulp and paper workers.[12]

In still other cases, state and city AFL officials offered crucial help as restive workers began the process of organization and then sought guidance. In the South, local unionists often found rep-

11. The Brotherhood files for 1933 and 1934 are filled with reports of organizing activity around the country. For Wisconsin, e.g., see the files for organizers Jacob Stephan and R.C. Henke, as well as files for Locals 65 (Green Bay), 81 (Appleton), and others; for the Pacific Northwest, files for such newly established local unions as 169 (Hoquiam, Wash.), 153 (Longview, Wash.), 155 (Port Angeles, Wash.), and others; for New England and New York State locals, see files for such locals as 117 (Norwood/Oakfield, N.Y.), 4 (Palmer, N.Y.), and 80 (Oldtown, Maine) and the files for Vice-President Sullivan and international auditor George Brooks.

12. T.G. Keirn to Burke, July 30, 1933; William L. Connolly, President, Pawtucket Typographical Union No. 212, to Burke, Oct. 3, 1933; H.M. Grimes to Burke, Sept. 1, 1933, Reel 2; Floyd B. Rush to Burke, April 13, 1934, Reel 3; and S.W. Justus, Financial Secretary, International Brotherhood of Boilermakers, Clifton Forge, Va., local, to Burke, July 27, 1933, Reel 3.

resentatives of the International Brotherhood of Paper Makers on the scene and worked through them to create locals of the Pulp, Sulphite, and Paper Mill Workers, while in Wisconsin the two unions' helping roles were often reversed. Throughout 1933 and 1934, the Brotherhood seized any available means in local situations, regardless of jurisdiction, following leads supplied by representatives of the AFL or of other international unions as they chartered scores of new locals.

During all of this disorderly organizing, Burke and his staff found the presence of new recruits who served as special organizers both exhilarating and troubling. Perceiving the paucity of skilled organizers and keenly aware of the union's desire to seize upon the opportunities now so suddenly available, activists hoped to find careers in the labor movement while they fought the good fight for their fellow workers. Often the president-secretary had no choice but to put men and women on the payroll without inquiring too closely as to their backgrounds or qualifications. Occasionally, these hastily appointed organizers turned out to be charlatans, drunks, or petty embezzlers.[13]

For union officials, however, these disheartening experiences were more than counterbalanced by success stories among new cadres. Rasmus Anderson, a local Wisconsin activist, became a special organizer and played an important role in establishing the union in that state. And in Akron, Edward Mangan in 1933 began a long career in the union as a volunteer organizer. A blacklisted member of the Switchmen's union in the 1910s, he had talked union off and on for years but had received no encouragement from his co-workers. When Akron rubber workers began organizing into federal labor unions, however, Mangan renewed his efforts. He attended meetings and conferred with AFL organizer Paul Smith, who in turn directed him to the Akron Labor Temple, where he found good advice and willing help from Wilmer Tate and W.H. Wilson of the Central Labor Union. These AFL functionaries put him in touch with Burke and helped him to organize his own shop in Akron and to begin organization of a large boxboard plant in neighboring Rittman. He began organizing box shops in nearby Cleveland, partially on his own time, partially as a special organ-

13. George Clausen to Burke, Nov. 26, 1933, Reel 2, and May 8, 1934, Reel 2; Burke to A.B. Hoff, Jan. 30, 1936, Reel 2.

izer for the international union. After being elected president of his local in January 1934, he traveled to Washington to appear on behalf of box workers at NRA code hearings. His employed punished him for his union activism by putting him on half time, but undaunted he promised Burke that "I will not quit as I am either going to make this a union paper box workers town or go broke trying to."[14]

The variations in local organizing experience were endless. In Marrero, Louisiana, Peter Sears kept hearing co-workers talk about unions. Knowing nothing about labor organizations, he attended a mass meeting, heard some persuasive oratory, and joined up. Not one to take a back seat, Sears appointed himself an unofficial organizer. "One pay day in Aug. [1933] . . . I [stood] at the gate. Collected thirty dollars. At twenty five cents a member." Alas, the enthusiasm among the ranks failed to transform these men into regulars, and by January 1934 Sears and one other worker remained the only union men left at the mill. In entirely different circumstances, left-leaning David Gordon of Local 107 Paper Plate Makers in New York City described his efforts to recruit fellow workers. The son of Local 107's president, Hyman Gordon, a vocal supporter of radical causes, David Gordon explained that the organizer had to be careful not to frighten workers, even in sophisticated New York City. As he sought to expand Local 107 into paper novelty shops in the city, he learned to rely upon a few key individuals in each plant. Workers had to be prodded gently, not harangued. Some workers remained suspicious, viewing all unions as racketeering enterprises or seeing Gordon as an outside troublemaker. If workers wanted to try the company union route, Gordon advised the organizer to accept their decision, gradually bringing to their attention the limitations of this form of "organization." At one shop, workers opted first for a company union, "known as the Brooklyn Standard Boys' and Girls' Benevolent Association," but after Gordon had spent a month patiently ex-

14. Rasmus Anderson, "The Problems of the Organizer," Brotherhood *Journal* 22 (Jan.-Feb. 1938), 3; Brotherhood, *Report of Proceedings . . . 20th Convention* (1944), 36–37; John P. Burke, "Tribute to Rasmus Anderson, Brotherhood *Journal* 28 (May-June 1944), 2–4; Mangan to Burke, July 30, Aug. 2, Aug. 24, Aug. 27, Sept. 16, Oct. 15, Oct. 22, Nov. 13, 1933, Reel 1; and Mangan to Brotherhood *Journal*, Jan. 20, 1934, Reel 1. The words quoted appear in the Nov. 13, 1933, letter. See also Mangan to Brotherhood *Journal* 26 (July-Aug. 1942), 5.

plaining its drawbacks, they discarded this plan "and came into the Union."[15]

Whatever their method of organization, new recruits expected immediate benefits. Most of the new members had had no previous experience with organized labor. They regarded their initiation fees and monthly dues payments as investments from which quick dividends were to be expected. If union membership did not result in prompt wage improvements, interest rapidly declined. Attendance at meetings slackened sharply. More importantly from the international's viewpoint, dues collections fell off and in many cases stopped altogether.[16]

From the point of view of the ordinary pulp, paper, or box worker, such a reaction was entirely justified. In the early stages of organization, supporters of the union had emphasized the direct and tangible benefits that would accrue to workers who joined the union. Burke, in high-minded editorials in the *Journal*, stressed the broader purposes of organized labor, invoking the spirits of Eugene Debs, Mother Jones, and other labor martyrs. The organizers, however, appealed to workers on grounds of self-interest. Enthusiastic promises sometimes boomeranged. At Kimberly-Clark mills in Wisconsin, antiunion workers explained to Brotherhood representatives that they enjoyed higher rates of pay and better working conditions and benefits under the company union than did organized pulp and paper workers in the region. As a result they of course saw no reason to abandon their current organization, which required no dues, to embrace the bona fide unions. Unionists could only reply that in the long run, Kimberly-Clark would no doubt chisel on wages and conditions, leaving workers who had not formed their own organizations with no redress.[17]

For their part, long-time supporters and local activists who had committed themselves to the Brotherhood resented the behavior of the new members. One Toledo unionist complained, "Be-

---

15. Peter Sears, Local 149, to Burke, Jan. 23, 1934, Reel 3; Gordon in Brotherhood *Journal* 18 (April 1934), 7.

16. E.g., C.L. Henderson, Local 79, Brainerd, Minn., to Burke, April 9, 1934, Reel 2; Burke to C.B. Werder, Local 156, Port Huron, Mich., May 6, 1934, Reel 2; Mrs. Loretta Thompson, Local 197, Cincinnati, to Burke, Dec. 20, 1934, Reel 2.

17. E.R. Henke, Local 81, Appleton, Wisc., to Burke, Aug. 1, Oct. 5, 1933, Reel 1.

fore you can get people to pay dues they are almost as bad as bargain counter hunters. . . . they want value rec'd in advance." Sounding very much like a veteran international functionary, Edward Mangan said of new unionists whom he had organized in the Akron-Cleveland area, "They think all they need to do is join an organization one day and the organizer can go to employers the next day and make demands and get them."[18]

Long-term international representatives were even more critical, castigating new unionists for impatience, lack of loyalty, and downright fickleness. Acerbic George Brooks reported from New England and from the upper South, contemptuously dismissing many of his new charges for expecting so much in return for the most minimal of commitments to the organization. Workers toiled for years at starvation wages, he sneered, without complaint; the moment they parted with one dollar in union dues, they expected immediate and substantial recompense. From around the country, local union officers castigated their fellow workers for apathy and dues arrearage. Leo Bujold of Local 25 in Rumford, Maine, an old newsprint local, reported that more than 250 members were at least three months behind in dues and that membership was steadily dropping. John Smole, secretary of Akron Local 145, called his co-workers "stubborn and dumb . . . , the most jell[y] fish bunch of weaklings I ever seen and yellow."[19]

These reports of inconstancy and defection confirmed Burke's judgment of American workers. Pugnacious and militant one moment, they lapsed into torpor and indifference the next. "I find everywhere among the workers that just as soon as they meet with some little obstacle they immediately become discouraged and quit,"[20] he observed to a unionist from Green Bay, Wisconsin. For the president-secretary, unionism was a sacred trust, a highminded effort to advance the general good. Having been a union member since 1905 and an officer since 1906, Burke had no sympathy for the narrow calculus upon which new recruits based their support.

Union leaders, both local and international, recognized that

18. Lambert Louy to Burke, Jan. 16, 1934, Reel 4; Mangan to Burke, April 13, 1934, Reel 1.

19. Leo Bujold to Burke, April 16, 1934, Reel 4; John Smole to Burke, Sept. 7, 1934, and c. Dec. 17, 1934, Reel 2; Brooks to Burke, Dec. 14, 1934, Reel 1.

20. Burke to Alex Schoen, Local 65, Dec. 22, 1934, Reel 3.

however unjust or unrealistic they might believe the ranks to be, signed agreements with employers incorporating wage increases were central to the organization's hopes of success. In his many appearances at code hearings in Washington in 1933 and 1934, Burke cultivated cordial relations with major company representatives. He went out of his way to stress that his stance, also that of the AFL, was moderate. He made particular efforts to reopen communications with large employers and clusters of employers, aiming for eventual multiplant and even multifirm arrangements. He conferred in 1933 with Jacob Graustein of International Paper about the possibilities of reinstituting union shop agreements with the giant company. He cultivated Robert B. Wolf of the Weyerhaeuser Company in Longview, Washington, long a friend of organized labor and an influential manager of a West Coast pulp mill. For Burke, the expansion of unionism in the industry was a twofold process: it required the creation of stable, widely supported unions on the local level and expert negotiations with shrewd but reasonable employers at the national or regional level. The result, he believed, would be a stable, progressive industry. Labor would be well compensated, the union would be self-sustaining and established on a firm footing, and employers would enjoy orderly, productive labor relations.[21]

Apart from the difficulties experienced by the international union as it sought to keep local unionists in line, Burke's program for the revitalization of collective bargaining in the industry met with three major obstacles. In the first place, converted paper did not fit into his scheme, for it was a decentralized industry requiring hundreds of local bargaining arrangements. Second, however reasonable and well informed he found some of the company officials with whom he conferred, the pulp and paper industry undeniably had its share of antiunion employers who could be relied upon to fight organization with every resource at their disposal. Finally, the NRA itself, however much Section 7(a) encouraged union organization, distorted the establishment of collective bargaining and complicated the union's tasks of organizing and bargaining.

21. Graham, *Paper Rebellion*, 14–15; Burke to James A. Taylor, President, Washington State Federation of Labor, June 29, 1933, Reel 3; Burke to William Green, June 27, 1933, Reel 3; Robert Wolf to Roderick Olzendam, Aug. 4, 1934, copy in Weyerhaeuser file, Brotherhood Papers, Reel 6; Wolf to Burke, Sept. 18, 1934, Brotherhood Papers.

Burke welcomed the opportunity to extend the union massively into converted paper. He hailed the creation of local unions in Cleveland, Toledo, Akron, and Rittman, Ohio, and in Milwaukee, Wisconsin, as the opening wedge "in organizing this large field."[22] There were thousands of paper box, bag, and novelty shops around the country. Even with the expanded staff and the participation of local activists as special organizers, the international union found it difficult to maintain the local organizations and to initiate negotiations with employers who were, as a whole, extremely suspicious of organized labor and entirely willing to employ any means necessary to purge their workers of the union virus. Paper box and bag workers felt that they had demonstrated courage by simply signing up with the union. Having joined up and paid their dues, both figuratively and literally, they now expected strong and sustained support from the organization. When employers stalled, fired union activists, or sought to intimidate workers, the new unionists expected the Brotherhood to fulfill the promise that in union there was strength: the president-secretary should rush skilled representatives to their aid.

For his part, Burke found it impossible to deal with the disputes and grievances of dozens of new converted paper locals. He was constantly asked to dispatch veteran organizers; members argued that their own local representatives were too inexperienced and too little respected by management to negotiate forcefully. If the Brotherhood did not respond immediately, local unionists complained of fraud and dropped out or defected to a company union.

But converted paper was too vast a field to ignore. Burke and his aides remained committed to organizing their jurisdiction and to working on behalf of converted paper workers despite frequent feelings of frustration and dismay. In December 1934 the president-secretary reflected, "We have made some progress in organizing these converting plants, but have awful problems on our hands every time we organize these workers," for "they require the constant attention of an organizer."[23]

In addition to the problems associated with the inexperienced and impatient recruits, the international union encountered scores

22. *Brotherhood Journal* 18 (Oct. 1934), 1.
23. Burke to George Brooks, Dec. 13, 1934; Zieger, "Limits of Militancy," 646–50.

of antagonistic employers. Hostility toward unions in primary pulp and paper went back a half century. For every "reasonable" official with whom Burke and his fellow officers conferred amicably at code hearings, there was another to whom the union was anathema. From every area of union activity, reports flowed in of harassment, intimidation, evasion, and manipulation of NRA code regulations. Many employers regarded unions as simply irrelevant: since economic laws dictated how much money was available for wages and benefits, a dues-consuming outside organization was at best an annoyance, at worst a malignant influence that would stir up discontent to justify its existence.

Because many employees believed that the NRA required some sort of employee representation, or at least that the establishment of an employees' group helped further President Roosevelt's aims, employers scrambled to establish unions, providing meeting space, meeting time during working hours, and, most important, financial support. Company unions flourished like mushrooms after a spring rain throughout the jurisdiction of the Brotherhood, competing with the bona fide unions for the workers' allegiance.[24]

Harassment, firings, and discrimination were rife in all sectors of the industry. The tenuousness of the union's reorganization, the inexperience of many new members, and the cumbersome procedures of the federal agencies set up under the NRA created difficult problems for organizers. In July 1933 a Central Labor Union official in Tuscaloosa, Alabama, who was seeking to organize about 600 kraft mill workers at the Gulf States Paper Corporation, described the company's methods to Burke. These workers, said F.W. Wenth, know nothing of organized labor, but they "want to get organized." The company had, he added, discharged several union activists, but "morale at present is good and the fireing [sic] of these employees has set them on fire." It seemed essential to "force the company to put them back to work," a tall order for so tentatively

---

24. Gladys Funkhauser, Secretary, Local 169, Hoquiam, Wash., to Burke, Sept. 26, 1933; Homer Clark to Burke, April 10, 1934, Reel 3; J. Le Fleur (?) to Gus Ramaker (Local 157, Tuscaloosa, Ala., file), early 1934, Reel 4. The efforts of the West Virginia Pulp and Paper Company to establish a company union while carefully avoiding public relations and legal problems are revealed in extensive intracompany correspondence, 1933–35. E.g., David L. Luke, Jr., to James Towsen, July 10, 1933, Westvaco Papers, Department of Manuscripts and Archives, Cornell University Libraries, Box 39. See Boxes 39–40, passim.

organized a local and for its international union, which at the time employed only five field representatives in the entire country.[25]

Firms fought back with company unions and firings, and some of them found it advisable simply to string out conferences until union members became disgruntled and withdrew their support. R.J. Sund, director of manufacturing at Marathon Mills in Wausau, Wisconsin, advised a fellow operator in a nearby town to stall negotiations as long as necessary, as his company had done earlier. "The employees," he counseled, "are quick to lose interest because in joining the union their natural expectancy is that they are going to secure some immediate benefits." When gains are not quickly forthcoming, "they rapidly start dropping out of the union. I cannot help but feel that if you follow such a stalling procedure . . . this thing will wash out."[26] At a mill in Covington, Virginia, management combined sponsorship of a company union with delaying tactics. "Of course it has gone against the grain to sit back and let . . . fellows go ahead and sign up some of our fellows in a labor union," confessed the plant manager. But he believed that in the long run the avoidance of outright opposition, together with vigorous sponsorship of a company union, extended conferences in lieu of negotiations, and close scrutiny of the whole process, would prove successful and would not needlessly arouse the workers.[27]

The very fact of the NRA with its boards, codes, and procedures added to the organizational problems. To be sure, the establishment of the NRA and the promise of Section 7(a) were crucial to the union's resurgence, but at the same time, the new agency's activities often frustrated and sometimes harmed workers and organizers alike. Employers' associations dominated the code-making process, with union representatives relegated to the background. Administration of the codes likewise depended on the good faith and cooperation of pulp, paper, and converted paper manufacturers. In some segments of the industry, code violations often involved chiseling on wages and labor conditions. Most problematic, however, was the NRA's apparent inability to induce employ-

25. F.W. Wenth to Burke, July 31, Aug. 12, Aug. 14, 1933, Reel 3.

26. Sund to manager of Rothschild, Wisc., mill, July 1, 1935, Marathon Paper Company Records, State Historical Society of Wisconsin, Reel 1.

27. George Miller to Thomas Luke, Sept. 6, 1933, Westvaco Papers, Box 39.

ers to take organized labor's rights seriously and to create effective machinery to deal with the recurrent labor disputes.[28]

Over the years the Pulp, Sulphite, and Paper Mill Workers had vigorously endorsed the voluntaristic traditions of the American Federation of Labor. Although Burke and other officers had at times participated in socialist politics, and although some staff representatives had held public office, the union itself had shied away from overt political activities. Nor had union leaders ever sought a strong governmental presence in the industry, confining their few contacts with officials over the years to occasional testimony before legislative bodies on conservation, pollution, and tariffs. Moreover, Burke and his closest associates believed that free and independent trade unions remained the workers' best hope of achieving justice. Brotherhood veterans did not welcome the intrusion of government into economic life and were especially suspicious of its entry into the very special environment of collective bargaining.

Nonetheless, unionists had no illusions about the importance of NRA and its labor provisions in the resurgence of their organization. Throughout the industry, union representatives, whatever their private doubts, aligned themselves with the NRA. Under its provisions, Burke maintained, "the unions are given a wonderful amount of recognition." From Wisconsin, Jacob Stephan reported intense interest among workers in the code-making process and urged workers to fulfill their duty to the administration by organizing into bona fide unions. Paper workers eagerly participated in NRA parades and demonstrations. Local 145 in Akron "turned out nearly 100% and all the girls carried a box" in a parade in October, according to the local's president. Clearly the international union could not afford to indulge its voluntaristic ideology, given the widespread support for the NRA among potential members.[29]

At the same time, unionists of long standing detected problems with NRA-inspired unionism. They resented the fact that a legislative enactment filled their ranks when decades of their own efforts had not done so. Years later, Burke became angry at sug-

28. Burke to Matthew Burns, July 3, 1941, Reel 3.
29. Burke to Brooks, June 10, 1933, Reel 1. Stephan in Brotherhood *Journal* 17 (Sept. 1933), 3; Mangan (from Akron) to Burke, Oct. 15, 1933, Reel 1.

gestions that the NRA had brought salvation. "I think it is a mistake . . . to give the impression that before 1933 the unions were weak and struggling and ineffective," he rebuked the editor of a government-produced history of labor relations. "So far as this union is concerned it has been an important factor in the pulp and paper industry for forty years."[30] Veteran organizers could not always suppress their contempt for the members now signing up—workers who, in their view, had abandoned or ignored the organization during its days of tribulation but who, now that there were no risks, flocked into it expecting quick results. Thus in November Vice-President Sullivan hoped that some recent improvements in the NRA paper industry code would boost the organizing campaign in Maine, but he wondered "if the dumbbells can think any more." From Fall River, Massachusetts, George Brooks reported in May 1934, "It is not worthwhile to waste any time on those corrugated and hat box workers as they show no desire to organize despite the fact that their employers are chiseling them plenty." Long-time staff members frequently referred to the new recruits as children, easily excited but easily discouraged, unlikely material for the Brotherhood's ongoing work, always needing tutelage in the practice of unionism.[31]

The leaders of the union also feared that the volatile neophytes would misread the lesson of the NRA. Too often it seemed that workers had gotten the notion that any gains they had made had been bestowed upon them by the government. If wage improvements were not forthcoming, the union was blamed; if benefits were received, the NRA was responsible. Representatives hammered home the argument that Section 7(a) would not have been possible without strong union representation before Congress, that good codes and fair administration of them would require constant vigilance on the part of strongly supported unions, and that, in the final analysis, workers had to rely on their own organized strength to keep government and business in line. Still, union officials often felt that theirs was a losing cause. The more workers organized before the codes were written the better, remarked George Brooks, since he feared that "after wages and working con-

30. Burke to Boris Stern, March 10, 1941, Reel 8.
31. Sullivan to Burke, Nov. 14, 1933, Reel 1; Brooks to Burke, May 15, 1934, Reel 1.

ditions are set by government order it will become a harder job to convince brother worker that he needs the union." Burke grew alarmed at reports that some employers had interpreted NRA provisions as meaning that they could no longer agree to union shop clauses in union contracts. "If this is true," he said, "I am afraid that the . . . law will do more harm to our unions than good, because . . . only very few workers will stay in the unions voluntarily."[32]

However ambivalent their feelings about the NRA, Burke and his colleagues eagerly pursued opportunities to participate in code proceedings. They were skeptical about the fairness of the code-making process, but they nevertheless identified several benefits that could redound from careful presentations at hearings and from their service as labor advisers. They did of course hope to shape the codes in a way that would help workers in the paper industry; in addition, however, the proceedings offered an opportunity for productive meetings with employers on neutral ground. When Burke and Sullivan appeared before the government bodies, they could stress the Brotherhood's commitment to stable labor relations and their desire to act in harmony with leading companies. On such occasions Burke could renew acquaintances with employers and for the first time could make contact with influential Pacific coast executives. Code proceedings further served the purpose of reminding unorganized and newly organizing workers of the need for a union presence in the industry. As Burke repeatedly told workers, bargaining with employers and representation of workers in the shops and mills constituted only a part of the Brotherhood's responsibilities under the NRA; his own appearance at code proceedings entailed a major commitment of the organization's resources.

The same point had to be made to government functionaries and to employers. Industry dominated the code-writing process. Only the most efficient and articulate unions could begin to claim the attention of the authorities. Employers in the pulp and paper industry were unusually well prepared; trade associations in their industry had origins far back in the nineteenth century. Large sections of the paper industry codes were simply copied from rec-

---

32. Brooks to Burke, July 26, 1933 (first quotation) and July 31, 1933, Reel 1; Burke to Stephan, Nov. 14, 1933, Reel 1 (second quotation).

ommendations made by the American Pulp and Paper Association (APPA). Moreover, the very nature of the industry, with its
high level of technical and engineering expertise, ensured employers of access to reliable, articulate, and energetic spokesmen in dealings with the government. Government administrators learned
to respect and trust paper industry representatives and credited
the APPA with candor and frankness.[33]

Labor had far less prestige. The Paper Makers and the Pulp, Sulphite, and Paper Mill Workers could claim only sporadic organization of the industry, even after the wave of unionization in 1933
and 1934. Moreover, the entire structure of the NRA militated
against equality of representation for labor or consumer groups.
The Labor Advisory Board (LAB), established in part to rectify this
situation, combined powerlessness within the councils of the NRA
with lack of regard for the representatives of organized labor who
functioned as labor advisers or who participated in code deliberations. LAB administrator Gustav Peck accused AFL representatives
of failure to cooperate with the board and of using its deliberations to further their organizations' narrow purposes. "In my opinion," declared another LAB staff member in October of 1933, "a
Labor Advisor from organized labor, who sits in on a single code,
is so green and lacking in insight . . . that he does his case more
harm than good." The proper role of organized labor's representatives, this NRA staff member felt, was "as window dressing."[34]
Workers' interests within the NRA, then, were to be handled by

33. R.A. Lawrence, "Preliminary Report on the Paper and Pulp Industry,"
March 1935, NRA Records, Administration, Research and Planning Division, Code
Administration Study, vol. 1, Records of Divisions, Records of Research and Planning Division. Not everyone in the Roosevelt administration agreed with this
assessment of the pulp and paper producers. Charles Wyzanski, Jr., solicitor of
the Department of Labor and an influential architect of the labor legislation
of the New Deal years, observed in November 1933, "I have had considerable
experience in dealing with the paper and pulp people and I know that they are
as lacking in a public spirited view as . . . any other reactionary industrial group.
I very much suspect that any wages to which they would agree would be far
below those which ought to be paid" (Wyzanski to Frances Perkins, Nov. 14, 1933,
Franklin D. Roosevelt Presidential Library, FDR Papers, Official File 466, Folder:
NRA Codes P and Q).

34. Gustav Peck to Leo Wolman, Oct. 25, 1933, enclosed in Peck to Wolman,
Dec. 11, 1933; R.M. Wilmotte to Peck, Oct. 25, 1933; A.H. Myers (quoted letter)
to Peck, Oct. 25, 1933, NRA Records, Labor Advisory Board Files, Wolman, E-T,
File No. 7: Labor, American Federation of.

LAB functionaries who wanted as much as possible to avoid consultation with the unions, with their parochial perspectives and their representatives' ignorance of the complex code-building and administering process.

Throughout the fall and into the spring of 1934, Burke and Sullivan journeyed repeatedly to Washington, sometimes to present testimony, sometimes as NRA labor advisers. In their depositions, the unionists attempted both to serve as champions of pulp and paper workers and to persuade government officials and paper industry representatives of their moderation and judiciousness. Burke, who made most of these appearances, maintained a public attitude of cheerful cooperation. He enlisted his union in the Blue Eagle effort, expressing determination to help make the recovery program work. He sought out employers with reputations for fair dealing and remarked that the industry in general was "not antagonistic toward labor." At the code hearing for the pulp and paper industry, he listened as labor adviser while representative of the company union at Kimberly-Clark proposed a fifty-cent-an-hour minimum wage and a six-hour day for the industry. His criticism of these ambitious goals persuaded West Coast employers attending the session that they should rethink their company union schemes and should seriously consider welcoming the paper unions into their plants in Washington and Oregon.[35]

Still, as the various codes emerged in pulp, paper, and converted paper, and especially as revisions were proposed and debated, Burke and Sullivan attacked both the process and the results. The unionists objected to wage rates, especially in the converted paper industry, where they felt that urban conditions prevailing in that sector required higher minima than the thirty-two and thirty-five cents an hour that employers favored. They also spoke out against wage differences among regions and between the sexes — matters, again, of particular concern in the converted paper field. "I cannot adequately emphasize the objection of labor to such discrimi-

35. Burke, Statements, in Hearings 405-1-04 (Paper and Pulp Industry — Proposed Revisions), Sept. 14, 1933 and others in NRA . . . Transcripts of Hearings. Telephone interview with George W. Brooks, former Director of Education and Research, Brotherhood, Feb. 7, 1979. This George Brooks should not be confused with the George C. Brooks, who served the union for many years as international auditor and whose correspondence is frequently cited in this book. They were unrelated.

nation," Sullivan told the Sanitary Milk Bottle Closure Industry hearing in November 1933. Sexual and regional differentials not only were unfair to the southern workers and women, who were directly affected, but also encouraged the movement of facilities to low-wage areas and the substitution of women for men. In their battle against discrimination, Burke and Sullivan were joined by representatives of the Women's Trade Union League.[36]

Of all the unionists' criticisms, however, the most pressing concerned their objection to the lowered take-home pay that the codes sometimes initially entailed. Workers asserted that because minimum wages were low and ceilings had been set on hours of employment, pay envelopes grew slimmer with the coming of the NRA. Administration statisticians challenged this contention, arguing that the NRA had had a dramatically favorable impact on both wages and hours worked throughout the paper industry. Still, because the traditional long hours were reduced by the NRA codes, workers often did receive less income than they had before. "Ever since November 27th," Burke told the Paper and Pulp Code Revision hearing in June 1934, "I have been busy trying to explain to the workers how an increase in wages has resulted in less money in the pay envelope. . . . I have finally had to tell them that that is an N.R.A. paradox." He added, "Now, we do not want . . . any more N.R.A. paradoxes."[37]

In public, Burke sought to mitigate his criticisms with a show of cooperation and an awareness of limitations. "I pride myself upon being an economic realist," he told the Paper and Pulp Code Hearing in mid-1934. "I never chase fantasies." More privately, however, he grew impatient with the whole code-making process and continued to press organization, looking to the day when the code-making program might collapse or when organized labor would

36. Sullivan, Statement, Hearing 1608-02B (Sanitary Milk Bottle Closure Industry), Nov. 29, 1933, NRA Records . . . , Transcripts of Hearings; Elisabeth Christman, Statement, Secretary-Treasurer, National Women's Trade Union League, June 29, 1934, in Hearing 405-1-04 (Paper and Pulp Code Hearing–Revised), NRA Records . . . , Transcripts of Hearings.

37. Burke, Statement, Hearing 405-1-04 (Paper and Pulp Code Hearing–Revised), June 29, 1934, NRA Records . . . , Transcripts of Hearings. Figures revealing decline in weekly earnings are in Jones, "Paper Industry Study," 213–14.

have to face the fact of its powerlessness in the bureaucratic world of government regulation. Thus, as early as November 1933, even as he expressed sentiments of moderation and pledged support for the NRA experiment, Burke voiced private doubts. "The employers write the codes," he informed a labor researcher. "The only chance labor has to struggle is to appear at the public hearing and make a lot of futile protests and other amendments that we know before hand will not be adopted." The low minimum wages that had thus far been set, Burke observed ruefully to publicist Robert Dunn, "tell the story of how effective labor's 'struggles' have been in getting these codes."[38]

To Burke, as to his fellow officers in the union, the lesson, as always, was clear. The NRA might be helpful in stimulating interest; the code hearings might be useful as forums and as meeting grounds. But in the final analysis, once again, there was no substitute for organizing.

Now, of course, Brotherhood representatives found that they had come full circle. They had been unable to organize the industry without the presence of NRA and its putative support for organization, and Section 7(a) was an enormous help. Yet the organizers constantly feared that workers would attribute any gains to government, not to the union. Here were dilemmas enough for union organizers and international leaders. Adding to the complexity of the situation was the machinery established under the NRA for the settlement of labor disputes. The National Labor Board was created on August 5, 1933. Its successor, the National Labor Relations Board, was established by Public Resolution No. 44 and Executive Order No. 6763 in June 1934.[39]

38. First quotation: Burke, Statement, Hearing 405-1-04, June 29, 1934, NRA Records . . . , Transcripts of Hearings. Second quotation: Burke to Robert Dunn, Nov. 14, 1933, Reel 3.

39. The best general accounts of the labor relations machinery under the NRA are in Bernstein, *New Deal Collective Bargaining Policy* and *Turbulent Years*, 172–216. Although the NRA staff in August 1934 recommended the creation of a separate labor board for "this major industry" like those established in the petroleum, textiles, and automobile industries (among others), none was ever established; Max Kossoris, "The Paper and Pulp Industry," 2d ed., Aug. 7, 1934, NRA Records, Records Maintained by the General Files Unit, Consolidated Reference Materials, Special Research and Planning Reports and Memoranda, Paper and Pulp Industry.

Like much else associated with the NRA, the labor dispute machinery exerted a varying and ambiguous effect on unionization. Certainly, experience in the pulp and paper industry emphatically confirmed the general experience of labor unions under NRA, namely that the alleged guarantees of Sections 7(a) and 7(b) were far from ironclad and that shrewd employers could manipulate and obfuscate matters in such a way as to discourage organization. At the same time, the various regional labor boards did in general exercise a broad prolabor influence when they were called upon. These boards lacked powers to enforce NIRA provisions and were scarcely able to work through the Labor Advisory Board to discipline recalcitrant employers by depriving them of code protection and of Blue Eagle benefits. Still, they at times managed to reinstate fired unionists, to further the unions' claims to recognition through representation elections, and to advise fledgling unionists of their rights and opportunities. In some industries, such as auto parts and rubber, the delay, legal wrangling, and ultimate feebleness of labor board proceedings helped substantially to discredit the NRA among workers and unionists. In pulp, paper, and converted paper, however, because the industry was geographically scattered and diverse, the labor boards absorbed some of the workers' anger and impatience and provided the unions with opportunities to regroup volatile constituencies, to isolate obdurate employers, and to avoid bitter confrontations, for which the Brotherhood was rarely prepared.

From a situation in Terre Haute, Indiana, it was apparent that board staff members often saw themselves as facilitating the labor movement. In the fall of 1934, examiner Robert Cowdrill, working out of the Indianapolis Regional Board, found a delegation of workers who called themselves the Paper Makers' Welfare Association at his office demanding a representation election for their plant, the Terre Haute Paper Mill. Cowdrill attempted to delay the election, feeling that the company union did not truly reflect the workers' views. Admitting that he could not legally forbid such an election, he nonetheless explained, "It is not our desire to hold an election that will enable this company union to get the upper hand." Indeed, Cowdrill seemed more concerned about the fate of bona fide labor organizations than did the secretary-treasurer of the Vigo County Central Labor Union, who was serving as a local NRA board member. Cowdrill repeatedly attempted to persuade this man that the paper unionists should be informed of the threat posed by the company union.[40]

Mostly, however, labor board hearings involving the paper industry were inconclusive affairs involving disputed discharges. Lacking powers of compulsion, the regional boards, when finding for the worker, sought with mixed success to persuade employers to reinstate the discharged individuals. Even when the boards succeeded, they rarely attempted to secure back pay for aggrieved unionists. Instead they relied on employers' desire to restore harmony in the plant and to cooperate with the recovery program, patiently seeking to educate manufacturers to accept unions as legitimate, and even helpful, additions to their operations. When board representatives had a strong background in labor work, they could be vigorous and persuasive advocates. Often, however, they identified with the employers' values and priorities, showing little sympathy for the inchoate anger and frustration that workers displayed in their often risky efforts to establish unions.

Employers also took advantage of the regional boards' lack of funding. When labor board representatives visited the West Virginia Pulp and Paper Company's facilities in Covington, Virginia, for example, they were transported, fed, and housed at company expense and spent a great deal of their time chatting amiably with the articulate men who managed the plant. How much easier it was to understand the perspective of these college-educated engineers, managers, and personnel men than to listen to the inarticulate grievances of fledgling unionists, most of whom felt themselves jeopardized by their work for the union!

In another case, when New Orleans regional board examiners proposed investigating a labor-management dispute at the George & Sherrard Paper Company plant in Camden, Arkansas, local company officials protested that all decisions were made at the company's New York headquarters. The labor board budget did not contain funds for a trip to New York, nor did the board have subpoena powers to bring the appropriate executives to Camden. The problem was solved when George & Sherrard offered to pay

40. Correspondence between Robert Cowdrill and J.C. Prechtel, Oct. 2–14, 1934, National Labor Board-National Labor Relations Board (NLB-NLRB) Records (RG 25), NLB, Indianapolis Reg., Case 341, National Archives and Records Service, Suitland, Md. The quotation comes from Cowdrill to Prechtel, Oct. 9, 1934.

travel expenses so that board officials could visit company headquarters in New York. Unionists at Camden might have been forgiven a sense that the cards were stacked against them had they
been privy to the exchange of telegrams between the secretary of
the New Orleans board and George & Sherrard officials. "Will
leave for New York as soon as possible," reported the government's
man. "Would appreciate advanced expense." The company replied
the same day. "How much advance shall we wire you?" When board
members did visit Camden, they found a company car waiting
at the train station for their use during their visit.[41]

Despite the labor boards' weakness and inconsistency, they
played an important positive role in Brotherhood plans to gain
a foothold in the industry. Burke and his associates pursued a
dual policy of energetic organizing and cultivation of major employers. The labor boards served as a buffer between the union
and restive members in geographical and other areas of the industry in which the union's presence was tenuous and its ability
to press its program limited.

Relatively few labor board cases stemmed from activities in New
England and New York State or from efforts to organize the Pacific Northwest. On the other hand, several important labor board
cases resulted from efforts to crack antiunion strongholds in Wisconsin. In the South and in the converted paper field, cases involving discrimination and representation were frequent. Such
appeals to the labor boards reflected the union's grand strategy.
In old production areas, the Brotherhood sought to resurrect its
old locals and to reestablish union shop agreements with familiar employers.

In the Pacific Northwest, the paper unions attempted to convince large employers that union organization offered advantages,
working with sympathetic businessmen in the case of enterprises

41. George Miller, Manager, West Virginia Pulp and Paper Company, Covington, Va., plant, to David L. Luke, Jr., Jan. 27, 1934, Westvaco Papers, Box 39,
in which Miller observes that if only the NLB field man would "come over to
New York and sit in a meeting with you folks and take lunch with your Dad
[the Lukes owned the company] there is no doubt at all that he will find out
the clase [sic] of company we have, which will go a long way if we should be
called up before the Labor Board in Washington." About the Arkansas situation, see Logan-Young telegram exchange, May 21, 1934, NLB-NLRB Records, NLB,
New Orleans Regional Labor Board, Case 61.

such as Crown-Zellerbach and Weyerhaeuser. Aggressive pressing of labor board cases would have been counterproductive; the unions sought to appear reasonable and moderate, not contentious and combative.

In the South and in converted paper, the labor boards served the union's purposes effectively. The Brotherhood was in no position to organize these sectors thoroughly. It had too little money, too few organizers, and too little familiarity with conditions. At the same time, both areas offered possibilities for future growth. Confronting volatile situations and employers who were notoriously antilabor, hampered by a lack of organizers, and unwilling to ignore or abandon workers desiring organization, the union found that the labor boards stood between obdurate employers and ill-organized but often angry and resentful workers. Firings and discrimination cases could be channeled into the boards' cumbersome processes, sometimes defusing potential confrontations simply by delaying them.

In the converted paper industry, as well as in the South, newly organizing workers had very often had no previous contact with unions. "You can not imagine the fear the employees of this plant have of loosing [sic] their jobs," wrote Burke's representative in Louisiana in October 1933. Throughout Ohio and Indiana, paper box and converting plant workers encountered stiff opposition from employers, many of whom resorted to demotions, firings, and petty harassment.[42]

Such discrimination cases were potentially explosive for the union. Labor's ability to protect its most exposed activists and local organizers determined its credibility within the shops. Few issues could arouse workers more quickly and more angrily than the dismissal or demotion of a popular local leader. In Parkersburg, West Virginia, the Ideal Corrugated Box Company suddenly fired eight activists. Local unionists appealed to the local NRA board and gained reinstatement of seven of the people dismissed. But according to Walter E. Hoffman, financial secretary of new Local 164, "when the one was left out the boys got uneasy and said one

42. J.C. Furr to Burke, Oct. 16, 1933, Reel 2; Clyde E. Driver, Complaint of Violation of Code of Fair Competition for the Paper Box Manufacturing Code, July 18, 1934, NLB-NLRB Records, NLRB-I, Indianapolis Region, Docket 334 [also found under NLRB-2, Case C-656].

to another [, '] if we join that union we might get the same deal [']" and membership and interest plummeted.[43]

Apart from plunging into mass strikes, which international representatives viewed as enormously risky in terms of both local circumstances and the union's broader strategy, the Brotherhood could provide little direct aid for the aggrieved workers. Still, by pressing grievance and discrimination cases before the local and regional boards, international organizers and special representatives simultaneously demonstrated their willingness to support workers' rights, their knowledge of the law, their familiarity with administrative procedures, and their general ability to present and negotiate a case with the board and sometimes with otherwise reluctant employers. Both board functionaries and union organizers had an interest in preventing a wave of strikes in the industry and often cooperated with each other to defuse angry local confrontations between apprehensive and bitter workers and hardheaded employers.

The results of specific cases in the Middle West were mixed. Board intervention in Lockland, Ohio, led to arbitration and eventually to a union shop for Local 159. In Hartford City, Indiana, Jacob Stephan used the regional board to defuse a confrontation between an obdurate employer and restive workers, gaining the company's grudging agreement to post wage scales and to implement rudimentary seniority provisions while avoiding a strike that he believed could not be won. In York, Pennsylvania, however, Edward Mangan found than an employer who refused to cooperate with government representatives faced few penalties. "'What is the use of paying dues,'" he quoted angry workers as saying, when the employer disregarded the labor board and challenged the union directly. In this case, Mangan recommended, and Burke sanctioned, a strike. "We shall have to make plans to fight it to a finish," concluded the disappointed president-secretary.[44]

Although the Brotherhood's experience with the NRA labor

43. Walter E. Hoffman to Burke, Nov. 2, 1933, Reel 2.
44. Lockland: Arbitration Agreement, May 29, 1934, and related material, NLB-NLRB Records, NLRB-I, Indianapolis Region, Docket 206; Ella Mae Griffin to Burke, March 30 and April 14, 1934, Reel 3. Hartford City: Correspondence between Jacob Stephan and George T. Watson, July-Sept. 1934, NLRB-I, Indianapolis Region, Docket 334. York: Mangan to Burke, Feb. 25, 1935; Burke to Mangan, May 11, 1935, Reel 1.

machinery was mixed, in the final analysis union leaders found it more positive than negative. Auto workers might term the NRA the "National Run Around," and certainly paper workers encountered enough delay and disappointment to feel disenchanted with labor board procedures. Still, workers in the industry never entirely abandoned the view that the NRA was important to their welfare and an ally in their conflicts with employers.

Paper workers duly noted the 1935 Supreme Court decision striking down the NIRA. Unionists girded themselves for a "crucial and trying period" and fully expected employers to lower wages and to renew attacks on many an "N.R.A. baby," as new locals were often called. Burke warned organizers to play down the demise of the NIRA, noting the probable effect of the Court's action: "You know, it doesn't take very much to discourage the average working man."[45]

After the demise of the NIRA, Brotherhood members viewed the hectic activity of the preceding two years with mixed feelings. Membership had grown, reaching perhaps 15,000 by the middle of 1935. The union had largely succeeded in reorganizing New England and New York, becoming established on the Pacific Coast, and securing a foothold in Wisconsin. Still, claims of victory were premature. Indeed, Burke was unhappy about the results from two years of unprecedented activity. "The American workers had a wonderful opportunity to organize under the NRA," he declared, "but they refused to take advantage of [it]." Moreover, the gains that had been made were tenuous, for Section 7(a) of the act had provided the catalyst for organizing activity, and the labor boards had provided important help in the South and in converted paper firms. It remained doubtful as to whether the tender shoots of unionism could survive without government presence in the industry.[46]

Shortly after the demise of the NIRA, of course, Congress passed the National Labor Relations Act (NLRA). Clearly, organized labor would not long be without the protection of federal presence; indeed, the machinery included in the NLRA was designed in good part to eliminate the weakness and confusion that had crippled the labor boards under the NRA. Still, Burke continued to regret

45. Burke to Mangan, May 28, 1935, Reel 3.
46. Ibid.

the need to rely on government. His comment to a local unionist in October 1935, when the provisions of the new law were still unclear and labor's rights still subject to court determination, epitomized his attitude on the subject. "American workers," he counseled, "should not lean too much on the Wagner Labor Disputes Law or any other law. . . . The American wage earners must learn to organize and stay organized and then they can bargain collectively."[47]

47. Burke to C.H. Wensel, Oct. 22, 1935, Reel 3.

5

Organizing in the Thirties

THE DEMISE OF THE NRA DID NOT PROVE FATAL TO THE UNION'S efforts to organize the paper industry. Throughout the prewar period, the organization grew, surpassing the 60,000 membership mark in 1941. In this time of bitter labor strife and intense organizing activity, the Pulp, Sulphite, and Paper Mill Workers refined a cautious but persistent approach. Leaders and organizers took advantage of the new National Labor Relations Act and also used the U.S. Conciliation Service (USCS). In an age of sit-down strikes, mass picketing, and violent confrontations, the union resorted largely to quiet organizing campaigns, low-keyed negotiations, and the cultivation of cordial relations with employers. Officers and organizers crisscrossed the country by train and automobile to visit local unions, which had grown to more than 300 by 1941. Representatives adjusted grievances, negotiated with employers, and tutored paper workers in the quotidian details of trade unionism.

Suspicious of the mass activism associated with the CIO and repelled by the strong-arm tactics and coerciveness that were the stock in trade of such AFL affiliates as the Teamsters and Carpenters, Brotherhood officers stressed the tried and true virtues of organized labor. When critics found their approach to organizing and bargaining during these eventful years pallid and stodgy, union representatives proudly cited the gains made and the confrontations avoided. "I doubt," Burke boasted in 1940, "if there is any union in this country that has done so much for the low paid men as has our union"—and this achievement despite the chronic unwillingness of rank-and-file workers to provide sustained, voluntary support for the Brotherhood's ongoing work.[1]

Throughout the period of growth, the new federal presence in labor-management relations—the National Labor Relations Board, created by Congress on July 5, 1935—exerted a strong but often indirect influence in behalf of the union. Unlike the labor boards that functioned during the NRA period, the new board possessed powers of investigation, subpoena, and decision making. Although apprehensions regarding the law's constitutionality during its first two years blunted its immediate impact, the labor board certainly strengthened the morale and sometimes the actual status of the union. Certainly, President Burke regarded it as a friendly agency and, despite his loyalty to the Federation, publicly opposed efforts made by the AFL Executive Council in 1938 and 1939 to secure amendments to the law. In later years, exercising a somewhat selective memory, Burke and official union publicists hailed the law as the key to the union's success in the 1930s. "Taking immediate advantage of the protection given in the Act, the Union sent its Organizers into the hundreds of mill towns and big cities," declared an organizing manual of the 1950s.[2]

The truth was more prosaic. The union launched no sustained organizing drive immediately following passage of the Wagner Act, nor was the operation of the new law as dramatic as these words suggested. The uncertain constitutional status of the legislation during its first two years of existence encouraged employers to continue their discriminatory and dilatory tactics. Even after the Supreme Court validated the law in 1937, unionists often found the lengthy proceedings and legal maneuvers frustrating and disappointing.

A case involving Local 299, an organization embracing scores

1. Burke to J.B. Rooney, Sept. 1, 1940, Reel 4.
2. Burke, Editorial, Brotherhood *Journal* 23 (May-June 1939), 20–23; John P. Burke, untitled pamphlet (1954); Brotherhood, *Introducing Your Union*, rev. ed. (1954). Both pamphlets are in Vertical Files, Labor-Management Documentation Center, Catherwood Library, New York State School of Industrial and Labor Relations, Cornell University.

of small set-up box shops in New York City, illustrated the benefits and drawbacks that the union experienced under the new NLRB.
On the one hand, the local's members and business manager
found proceedings discouraging and time consuming. "Insofar as
workers in the industry are generally concerned . . . ," declared
the box workers' leader, "to organize and strive to better . . . conditions means to become a martyr." Discharged workers found
redress slow as discrimination cases bogged down in lengthy proceedings. Still, Local 299 relied heavily on the board in its efforts
to organize the fragmented and competitive set-up box industry.
Many employers were so intransigent and many workers so apprehensive that strikes to win recognition rarely succeeded, but
because employers eventually had to bow before the law, workers
concluded that federal power, however delayed in its impact, was
on their side.[3]

In Wisconsin, too, the new board was instrumental in breaking the open-shop bastions. The union won important victories
in representation elections and through board-issued cease-and-
desist orders in large mills in Wausau, Mosinee, and Nekoosa-Port
Edwards.

Victory in the latter Wisconsin River communities was particularly sweet. The Nekoosa-Edwards company, noted the NLRB
regional director, had "a prominent anti-labor record dating back
for a number of years" prior to the unions' breakthrough in 1938,
an achievement that brought in more than 1,000 new members
for the Pulp, Sulphite, and Paper Mill Workers and other AFL affiliates there. Late in 1941, Wisconsin organizer Raymond Richards revisited these mills, this time for a labor-management conference. "It all seems like a dream," he mused. "This meeting was
held in the same hall that I spoke in some years ago when I was
told that if I ever came out there again . . . I would be shipped

---

3. A.N. Weinberg to Charles Fahy, NLRB General Counsel, May 17, 1938; John
Maringolo to Ernest Gross, Jan. 5, 1939, National Labor Board-National Labor
Relations Board Records (RG 25), NLRB-2, New York Region [Informal Files], Case
C-227, National Archives and Records Service. Local NLRB officials agreed with
Weinberg's analysis in this case. "The Union's whole organization drive . . . in
the industry . . . [has] been hamstrung by the delay. . . . Little Weinberg . . . is
really pretty patient," Elinore M. Herrick informed the Washington office (Aug.
5, 1937). On the importance of the NLRB to Local 299, see Weinberg to Burke,
July 19, 1939, Reel 5.

out in a wooden overcoat." Another Wisconsin organizer declared in 1940, "The National Labor Relations Board has been of great help to our organization here in Wisconsin."[4]

Board proceedings could be risky. Occasionally an aggressive CIO challenger signed up enough workers to force an election, and the Pulp, Sulphite, and Paper Mill Workers found themselves outvoted, as in a large mill in Berlin, New Hampshire, in 1941. Even CIO challenges, however, did not dampen enthusiasm for the new law. Indeed, in some cases the Brotherhood found NLRB elections helpful in thwarting rivals. Otherwise recalcitrant employers, faced with inevitable organization, cooperated closely with the paper unions in the scheduling and arrangements for NLRB elections to deflect CIO initiatives. Thus, one Ohio company official, minutely informing Burke of CIO activities in his plant, advised the union leader to "have your men bring pressure on the NLRB. I will do the same!" Collusion between the union and company officials was particularly close in the organization of new mills in the South, as the United Mine Workers' District 50 and the CIO Southern Kraft Workers Organizing Committee offered sporadic challenges.[5]

As useful as the NLRB often was, the union relied primarily on its traditional ways of organizing and negotiating. Some participants in the labor movement felt that the times offered an opportunity for bold assertion and laborite transformation of society; for Burke, this was a priceless opportunity to establish the Brotherhood once and for all. The union had proven its steadfastness on scores of picket lines throughout its history; it had no need to storm the rhetorical barricades now. "I have been through a few Gethsemanes during the past twenty years," Burke told the union's 1937 convention, "but it has been worthwhile." All signs

4. Activities in Wisconsin are indicated in the correspondence and reports of Vice-President Raymond Richards, 1933–37, Reel 1 for each year. E.g., Richards to Burke, Jan. 7, April 10, 1935; Jan. 8, Feb. 15, Feb. 21, 1936; May 28, 1937. Also Richards to Burke, Dec. 2, 1941, Reel 1 (recalling earlier tribulations), and Rasmus Anderson, Monthly Report, March 1940, Reel 1 (quoted). For Mosinee, Case XII-R-19, Feb. 8, 1936–May 1, 1936, NLRB-2; for Nekoosa-Port Edwards, I have quoted Nathaniel S. Clark to Nathan Witt, Jan. 7, 1938, Case XII-C-598 (informal files) and Case XII-C-595 (informal files).

5. Burke to George C. Brooks, May 18, 1940, Reel 1; Rolland E. Friedman to Burke, Nov. 15, 1941, Reel 9; also James Towsen of West Virginia Pulp and Paper Co. to Burke, April 2, 17, 23, June 2, 1941, and related material, Reel 1, and Burke to T.A. McDonald, Jan. 12, 1941, Reel 9.

were favorable, he believed. "I do not know of any Union that has much brighter prospects for the future. . . . there is no reason why we cannot organize this entire industry during the next few years."[6]

Like all unions, the Pulp, Sulphite, and Paper Mill Workers on occasion resorted to the ultimate weapon, the strike. Burke favored ingratiating the union with employers and exploiting all means of negotiation and conciliation, but he believed that "the strike is labor's most potent weapon." While the organization had no single epic confrontation in the 1930s, strikes were almost always in progress somewhere. By 1941 the international union had accumulated a strike fund of more than $400,000. Always mindful of the rapid depletion of almost $250,000 in the strikes of the 1920s, Burke doled out strike benefits cautiously, following no specific union regulations. He attempted to limit financial support to strikes duly authorized — called, that is, only on recommendation of the organizer on the scene and with the approval of the international union's executive board — but he frequently relented and granted funds to locals that had acted precipitously and had ignored regular procedures.[7]

Strikes, in Burke's view, were risky enterprises. For every one that heroically galvanized workers into union stalwarts, a dozen ended in bitterness and recrimination. Of course, obdurate employers sometimes gave the union no choice. When a Bronx bag manufacturer challenged the union in 1936, Burke rose to the occasion: "Let us show this bird that when he tackled this International Union . . . he ran into an organization that knows how to fight." In Shelbyville, Indiana, the next year, bolstered by Burke's advice that he should teach an antiunion box manufacturer a lesson, Brotherhood Vice-President S.A. Stephens confronted the affronting capitalist. "He hit the desk and pounded it but every time he did . . . I done the same [and] . . . I told him I would pull his plants on strike."[8] More often, however, strike situations were

6. Brotherhood, *Report of Proceedings . . . 17th Convention* (1937), 46.

7. The size of the strike fund is given in George Brooks, Auditor's Report, 1941, Reel 1. Burke's disbursement practices are indicated in his letter to John M. Treier, June 29, 1937, Reel 7. I have quoted his letter to Clarence Jornlin, Jan. 18, 1941, Reel 4.

8. Burke to Jacob Stephan, March 27, 1937, Reel 2; Burke to S.A. Stephens, Aug. 19, 1937, and Stephens to Burke, Sept. 1, 1937, Reel 1.

ambiguous in Burke's view. Strikes resulted, he believed, when workers failed to maintain their organizations, thus encouraging employers to test their resolve. While an employer's actions might in theory be reprehensible, they were a natural response to a feeble union organization.

"It is strange," the president-secretary remarked, "that so many working men seem to think that a strike is some kind of a Sunday School picnic." Too often the heady enthusiasm and camaraderie of a strike's early days seemed quickly to degenerate into apathy and recrimination. In July 1936, an Ohio organizer supplied a picture of the jaunty picket line of Local 150 at the outset of its strike against the Ohio Boxboard Company in Rittman. Burke studied it skeptically. "Will these . . . be union men after the strike is over?" he asked. "The American workers must learn that there is more to this trade union movement than being heroic on the picket line."[9]

Rittman provided a particularly powerful example, in Burke's view, of the pitfalls of reliance on the strike. Since the local's organization in 1934, membership had fluctuated wildly, and dues paying had been sporadic. In the summer of 1936, Ohio Boxboard ceased to recognize the union, which by now had only a few members. When local workers proclaimed a strike, Burke had to choose between disavowing their action and permitting them to bypass international union channels. With more than 1,000 workers employed, Ohio Boxboard was a rich prize for the Brotherhood, and Burke eventually swallowed his anger with the workers and gave them substantial financial support. Still, he remained convinced that their own disorganization had emboldened management. Nor did the Ohioans endear themselves to the president-secretary when they twice turned down settlement plans that had been negotiated by Vice-President Sullivan and even subjected him to public embarrassment. "I cannot understand the kind of union men we have out in that section," fumed Burke when he was informed of his colleague's ill treatment.[10]

A union, in one of Burke's favorite figures of speech, was not

9. Burke to Edward Mangan, July 11, 1936, Reel 2.

10. Thomas P. Hyland, Summary of Final Report, Sept. 3, 1936, and related materials, USCS Records, File 182-1590; Burke, Correspondence, Summer 1936, Reel 2. Burke's statement appears in Burke to Mangan, Aug. 11, 1936, Reel 2.

a mob. Rather, it was a disciplined organization, cohesive and vigorous on the local level, alert and far-sighted on the national level.
Workers who hastily or foolishly went out on strike lost their independence and subjected themselves to control by outside parties. The officers of the Brotherhood, who were uniquely equipped
to see the overall position of the union, could not permit isolated
local groups to dictate the allocation of scarce union funds and
resources. "Boy, oh, boy," Burke once observed in exasperation,
"the boys and girls all want to get out on the picket line." As a
union leader, it was his task to channel and discipline this brave
but often misplaced enthusiasm.[11]

As its procedures involving strikes revealed, in the 1930s the
Brotherhood was not a very complex organization. Although the
number of organizers and local unions grew dramatically, the international union remained a simple operation, headquartered in
a small suite of rooms above a store in Fort Edward, New York.
At most nine or ten people, including Burke's wife, Bessie, and
her brother, Raymond Leon, handled correspondence, filing, the
collection of material for the *Journal*, preparations for the biennial convention, and other routine matters. The union had no
political arm, no research service, and no education or legal departments. When legal help was required, international representatives or local unionists employed local attorneys, although the
enormous legal expenses involved in the strike of 1921-1926 make
Burke extremely wary of involvement in legal proceedings.

Local unionists and international organizers received only on-
the-job training, with the *Journal* and occasional circular letters
from Burke serving to reinforce the international union's policy
of cautious negotiating, encouragement of worker's loyalty to the
company as well as to the union, and punctilious observance of
contractual provisions. Although some international representatives gained public office in the 1930s, the union itself remained
largely free of political endorsements and enthusiasms. Vice-
President John Sherman of Washington served several terms in
the state legislature and received union support, of course, but

11. Burke's correspondence with Local 150 Secretary Paul Smoyer through
the summer of 1936, Reel 3, indicates the financial arrangements through the
strike. Included in this file is an undated (but c. Sept. 1, 1936) financial report.
The words quoted appear in Burke to Sullivan, Oct. 10, 1936, Reel 1.

The Brotherhood's 1939 convention in St. Paul. In the front row from right to left: Vice-Presidents William Burnell, John Sherman, and Herbert W. Sullivan; President-Secretary Burke; Vice-Presidents Raymond Richards, Maurice La-Belle, S.A. Stephens, and Fred Morris; international auditor George C. Brooks. Also visible: Joseph Tonelli (circled, right) and Hyman Gordon (circled, center). Most of the women in the second row are wives of international officers. (Photo courtesy of Thurman Reynolds of Covington, Virginia, Local 152)

International headquarters of the Brotherhood at Fort Edward, New York, c. Labor Day, 1940 (*Brotherhood* Journal; *United Paperworkers' International Union*)

this mostly consisted of a place on the payroll and travel funds while he performed his legislative duties. The union also lent quiet support to Cong. George Schneider of Wisconsin, an officer in the Paper Makers, and Burke was proud of the fact that Vice-President Raymond Richards had been appointed to the prestigious Board of Regents of the university in that state. Aside from a very broad, even nominal, socialist orientation on the part of Burke and some of his closest associates, however, political interests absorbed little of the union functionaries' energies. Because the union represented widely diverse ethnic and regional groups, political appeals could only impede the process of organization. For the old socialist Burke, who voted for La Follette in 1924 and for Norman Thomas in 1940, "These so-called election campaigns between the Republicans and Democrats were just a big show to fool the people. There is no difference between them." [12]

Without an elaborate staff and without any complex system of regional or product suborganizations, the union in the 1930s functioned largely as the direct responsibility of John P. Burke. Six vice-presidents[13] served under him, and with him and Treasurer Bart Doody constituted the executive board. The board in turn served as the organization's decision-making body, with powers to sanction strikes, charter local unions, hire organizers, allocate funds, and approve contracts. Officers were elected on an at large basis at the convention. Burke was the only officer of the union, however, who was permanently stationed at its headquarters in Fort Edward. The vice-presidents and international auditor George Brooks functioned primarily as field representatives. Most of the work they performed in the field differed little from that assigned to organizers,[14] although, of course, Burke placed such veteran unionists as Brooks and Vice-Presidents Herbert W. Sullivan, S.A. Stephens, and Raymond Richards in charge of the most important and strategic campaigns and negotiations.

Since Burke appointed organizers (who were subject to nominal approval by the executive board), and since virtually all correspondence was directed to Fort Edward, the board rarely dissented

12. Brotherhood, *Constitution and By-Laws*; Burke to Morris Wray, Jan. 17, 1941, Reel 3.
13. A seventh vice-president was added by action of the 1935 convention.
14. *Constitution and By-Laws*.

from Burke's recommendations. On the other hand, the president-secretary was a modest and unassuming man, unimpressed with the trappings of office or with the delights of exercising power. Since veteran unionists held most of the vice-presidencies and since appointments for organizers emanated from Burke, the international bureaucracy, such as it was, ran on the basis of mutual agreement and consensus. Burke would not have considered pulling rank or imposing his will on his long-term associates, nor would they or the newer organizers contemplate rebelling against their respected president.

Still, dissent did sometimes occur. Impatient local unionists protested the international union's cautious organizing, negotiating, and grievance-handling practices. Unionists in the Pacific Northwest regularly expressed impatience and open disdain, dissidence that largely lacked broad political perspective and derived in good part from the sheer remoteness of their mills and union halls from Fort Edward and from the union's traditional areas of strength. At times, however, an articulate local leader would emerge, linking the Pacific Northwest's general grievances against the international union with enthusiasm for the CIO or protesting the inroads of AFL craft unions, about which the Brotherhood could do little. There were also occasional outbursts that recalled the region's IWW past. Pacific Northwest locals, together with many of the converted paper locals in Ohio, which frequently grumbled about the inadequacy of the international union's service and concern for them, regularly expressed their dissatisfaction with the international union at the biennial convention by opposing dues increases and supporting moves to reduce the per capita tax.[15]

The urban locals, particularly those in New York, manifested the only coherent ideological tendency in the union. Veteran organizers sent to service New York City locals were frequently perplexed by the ethnic diversity, ideological disputation, and sharp practices they encountered in the metropolis. In particular, Brotherhood veterans were wary of the political radicalism associated

15. Zieger, "Limits of Militancy," 649–51; Brotherhood, *Report of Proceedings . . . 16th Convention* (1935), 132–55, and *Report of Proceedings . . . 18th Convention* (1939), 152–70. Delegates from newly organizing areas supported dues reductions as an organizing device, arguing that aggressive recruitment required the Brotherhood to attract the uncommitted and that the resulting expansion of membership would actually increase the international union's revenues.

with Local 107 and its president, Hyman Gordon. Gordon and others in his family constituted a small corps of ideological left-ists in the union. Local 107, which represented paper plate and novelty workers in New York and Brooklyn, had been chartered in 1917 and was one of the Brotherhood's earliest converted paper locals. Gordon's sons and nephews circulated in the industry throughout the 1930s. His son David was elected president of a Toledo local, his son Lou served as an organizer in New York City, and another relative attempted to organize in New England.

Burke regularly sent the local $100 a month for organizing ex-penses but otherwise attempted to avoid and to deflect the left-wing statements emanating from Gordon and the local. The Brotherhood's constitution did not bar ideological leftists, and Burke did not censor communications to the *Journal* that came from the Gordons, who frequently reported on their attendance at radical rallies and meetings. On the other hand, the president-secretary did warn his organizers "not to bother Local 107 any more. Keep away from them. Let them have their little Communist heaven," thus imposing a kind of de facto isolation. For their part, the Gordons and their supporters in Local 107 sought to expand their operations in New York. They published one of the Brother-hood's first shop newspapers, sent members to help with other converted paper strikes and organizing campaigns, and spoke in favor of the CIO, industrial unionism, and international opposi-tion to war and fascism.[16]

If radical dissent did not rock the Brotherhood, the sudden pres-ence of large numbers of female, black, and ethnic workers did per-plex its leaders. The organization's movement into the converted paper industry, which had many of its facilities in large cities, greatly broadened the social base upon which union membership

16. The yearly file for Local 107 indicates the relationship between Burke and Gordon. The words quoted appear in Burke to Harriet Wray, Oct. 27, 1938, Reel 5. On the peregrinations of the Gordons: Burke to Dennis J. O'Connell, Local 168, Pawtucket, Jan. 15, 1936, Reel 3; David Gordon, President, Local 187, Toledo, Report, Brotherhood *Journal* 20 (Feb. 1936), 10; Burke to Jacob Stephan, April 23, 1936, Reel 2; David Gordon to Burke, July 28, 1936; and Burke to David Gor-don, July 30, 1936, Reel 3. For left-wing criticism of Burke in the Brotherhood *Journal* (the only such example), see comments of Isoda Drucker, Local 299, Brotherhood *Journal* 24 (July-Aug. 1940), 14–15. Burke's explicit opposition to giving the executive board the power to oust political undesirables is expressed in Burke to John Sherman, July 27, 1940, Reel 1.

rested. For thirty years, the Brotherhood had centered on New England, New York State, and Canada. Few women toiled in the pulp and paper mills, and ethnic diversity was provided mainly by outposts of French Canadians, who were an integral part of the union from its inception, and by the occasional groups of Irish and German Americans. But expansion in the 1930s brought the need to deal with a bewildering variety of urban paper and box workers, who included large numbers of union-minded women, and southerners, black and white. Brotherhood representatives, particularly in the early stages of the union's revival, reflected the traditions and values of the remote villages and small towns of the Northeast.

Of all the possible sources of perplexity associated with the new recruits, the presence of large numbers of female workers in the converting plants and box shops was the most obvious. The pulp and paper field had traditionally contained few women, but females composed more than 80 percent of the set-up box labor force in New York and other cities and from 20 to 50 percent of the workers in other shops producing bags, folding paper boxes, paper plates and cups, and novelties. More than 50,000 women workers fell within the Brotherhood's jurisdiction in the 1930s, and by 1941 more than 8,000 of them had joined the union. Women workers proved active and competent local unionists, particularly in the Middle West, where most of the converted paper locals had at least one female officer, usually the corresponding secretary.[17]

International representatives were ambivalent about these female workers. Some organizers were outspokenly antifemale. "Anything but girls in a strike," remarked old-timer George C. Brooks, "because for every one of them that has sense there are a hundred nuts." Local unionists often considered women workers in their midst to be a source of embarrassment or weakness. Paper companies, an Oregon unionist declared, should replace all fe-

17. The figures for female participation in the labor force and representation in the union are extrapolated from Auditor's Report, 1941, Reel 1. See also Brotherhood, *Report of Proceedings . . . 18th Convention* (1939), 1; [Baril and Jones], comps., "Wages . . . Set-Up Paper Box, 1933 . . . 1935"; Press Release, U.S. Department of Labor, Wage and Hour Division, "Converted Paper Products Committee Appointed," July 16, 1940, AFL Papers, State Historical Society of Wisconsin, Series 4, Box 97; and "Statement by Mr. H.W. Sullivan," Brotherhood *Journal* 24 (Nov.-Dec. 1940), 5-7.

male employees with men, a move that would obviate the need for federal relief and employment programs. Pittsburgh paper box workers reported that "women are always harder to organize than men simply because they figure they will be getting married in a few years so the H— with [the] Union." From Holyoke, Massachusetts, came the report that reorganization of Local 203 might succeed if some way could be found to deal with "the women members who made the Local the [scape]goat for their petty jealousies."[18]

Burke did not directly endorse such patronizing and disparaging sentiments, but he had no strongly articulated view of the industrial or social role of women. He tended to regard the issue as a matter of chivalry rather than justice. Particularly revealing was his response in 1941 to a local unionist who reported that the New Hampshire paper company for which he worked had instituted a highly restrictive employment policy regarding female workers. The Nashua Gummed and Coated Paper Company, as a member of Local 270 informed the union president, had decided to limit leaves of absence for pregnancy and maternity to six months; in the future, furthermore, it would employ no married women, and it would require prospective female employees to sign a pledge not to marry for at least a year after hiring. Burke was dismayed. He regarded the company's policy as unpatriotic: a nation that faced international crisis would clearly want to encourage population growth. "We should honor motherhood," he protested, adding, "The company has a right, of course, to hire single women in preference to married women," but he did disapprove of the agreement to postpone marriage: "I certainly think that any such rule is going pretty far in interfering with the personal rights of American women."[19]

The international union employed a few female organizers. Local activists in the Middle West were hired on a temporary basis to service nearby locals. New York's paper box workers' union contributed several female activists, some of whom the Brotherhood employed as organizers in the box and novelty shops of urban New England and the New York metropolitan area. In 1936, the

18. George C. Brooks to Burke, May 2 and 4, 1934, Reel 1; Earl Sharp, Local 165, Salem, Oreg., Report, Brotherhood *Journal* 22 (Jan.-Feb. 1938), 7; Local 286 to Keystone Set Up Box Company, Oct. 25, 1941, Reel 6; R.W. Jubinville to Burke, March 11, 1936, Reel 2. Also Zieger, "Oldtimers and Newcomers," 192.

19. A.E. Bartard to Burke, Jan. 16, 1941; Burke to Bartard, Jan. 18, 1941, Reel 6.

international office helped to send Valeria Brodzinski, from Local 201 in Menasha, Wisconsin, to Brookwood Labor College, from which she emerged as an able organizer and publicist for the union. But although Burke declared in 1937 that "an organization like ours that now has so many women and girls as members should have at least one young woman in this field as an organizer," Brodzinski's work remained sporadic.[20]

The international union made occasional efforts to develop a consistent approach to the problems of female workers. For a time, its representatives worked with Agnes Nestor and the Women's Trade Union League of Chicago in an effort to organize that city's converted paper industry, with its large contingent of women workers. Burke and his colleagues consistently supported the idea of equal pay for equal work, although they gave the notion little priority, especially after the demise of the NRA codes deprived them of a convenient forum for discussion of this subject.[21]

The ambivalence of the union toward women workers inevitably undermined the women's conception of their own role. Although women participated actively and effectively in a number of new locals, too often they felt discriminated against, rejected, or harassed by their male counterparts. Workers at a Toledo box shop, part of the multiplant Local 187, complained about the treatment they received from their male associates in other shops and requested a separate charter. Because they were women, they were subject to a much more rigid and stringent system of dues collection on the part of Local 187's male leadership than were other box workers. They were also discouraged from running for union offices. "We don't dare voice our opinion at the meetings," reported Mary Miller. The men came to the meetings drunk and argumentative, using vulgar language so as to intimidate and chasten the

---

20. I have quoted Burke to Valeria Brodzinski, Sept. 20, 1937, Reel 2. On Brodzinski, see her Report, Brotherhood *Journal* 20 (April 1936), 4; Brodzinski, Radio Address, Sept. 30, 1937 (WCFL, Chicago), with photo, printed in Brotherhood *Journal* 21 (Nov.-Dec. 1937), 13–14; and Brodzinski, "The First National Convention of the American Federation of Women's Auxiliaries of Labor," Brotherhood *Journal* 22 (July-Aug. 1938), 14–15.

21. Burke, Statement, Hearing 406-03 (Set Up Paper Box Industry), Oct. 3, 1933, NRA Records, Records Maintained by the Library Unit, Transcripts of Hearings; Agnes Nestor to Burke, Dec. 2, 1936, Reel 4; Burke to Nestor, Jan. 4, Jan. 12, 1937, Reel 7.

women workers, who naturally tended to avoid union meetings altogether, in the view of the treatment they received. Because their plant had been on strike and had received support from other shops, the men felt "that they should be able to tell us how to run our plant." The representative of the female workers on the local's joint negotiating committee was continually subjected to belittlement and ridicule. The embittered unionist asked her president-secretary, "Do you think that this is Brotherly Love and Unionism[?]"[22]

In the years prior to World War II, the Brotherhood did little to address these problems. Women continued to join as the union expanded even further into the converted paper sector. They continued to suffer serious disabilities in terms of wages and in terms of their treatment at the hands of fellow employees and union representatives. Dues for women stood at seventy-five cents a month; men were charged one dollar. In some urban locals, notably those in New York City, women took an active part in organizational and administrative work. To many male unionists, however, women remained an unassimilable element, a distraction — or in the words of George Brooks, "damned flappers."[23]

Veteran representatives also had a difficult time learning to organize workers of diverse ethnic and national backgrounds, particularly in urban settings. Burke and his long-time colleagues considered New York, Philadelphia, and other large cities alien territory, breeding grounds for corruption and radicalism. Moreover, the people associated with the labor movement in urban areas, often of exotic ethnic background, were frequently argumentative, assertive, and unconcerned with the Brotherhood's traditions of sober respectability.[24]

22. Mary Miller to Burke, April 15, 1941, Reel 7.
23. George C. Brooks to Burke, May 4, 1934, Reel 1. The by-laws prevailing throughout the 1930s required that the initiation fee be at least two dollars and that monthly per capita tax to be paid to the international union by the locals for each member be 60 cents for males and 40 cents for females. Although there was no specific amount for total monthly dues — to include the amount paid by each member for the support of the local over and above the payment to the international union — the standard rate was one dollar for men and 75 cents for women. See Burke to C.L. Ascher, Dec. 16, 1937, Reel 7.
24. Zieger, "Oldtimers and Newcomers," 192–94; Raymond Leon to Burke, Feb. 12, 1937, Reel 2.

Organizers viewed the cities as fraught with deceit and danger. Rumors of gangsterism, payoffs to public authorities, and infiltration by criminal elements retarded efforts to organize in northern New Jersey, while the rumor of collusion between Teamsters and converted paper employers in New York City and Philadelphia discouraged Pulp, Sulphite, and Paper Mill Workers' representatives on the East Coast. Jake Stephan reported on an encounter in 1936 with officials of a New York City box makers' union which had been organized under a federal charter but was in the process of entering the Brotherhood as a multiplant local. "You should be once here," he remarked to Burke, "and spend about two hours in the Box Makers office. I am sure you would be more than glad to get out. In all my life I have never heard so much comotion [sic], chewing the rag and pounding on the desk yelling at the top of their voice; in fact it is an insane asylum." The understanding Burke replied that the box makers talked incessantly of affiliation, but "do we want to get incarcerated in that madehouse [sic] again?" [25]

Pulp, Sulphite, and Paper Mill Workers' representatives were particularly suspicious of urban workers who evinced strong ethnic or national characteristics. Brooks declared that organizations among New England box workers were unreliable and unstable in good part because there were too many "Wops" (by which term he apparently meant anyone of southern European derivation). French Canadian paper box workers in Massachusetts, he believed, could not succeed in forming a viable local because of the inherent stupidity and laziness of these "Frogs." Other veteran organizers expressed their ethnic prejudices less overtly, but most found the multicultural world of the cities strange and difficult. In particular, Brotherhood representatives had difficulty relating to the Jewish and Italian workers of New York City and were frequently disconcerted by the volatility, loquaciousness, and general argumentativeness that they often perceived. By the later 1930s, however, after the absorption of the box makers' union by the international organization and after creation of several additional

25. On suspicions of Teamster kickbacks, see Morris Wray to Burke, July 14, 1937, Reel 3. Stephan's reactions to the box makers are contained in his letter to Burke, Aug. 18, 1936, Reel 2, while Burke's skepticism of the group is expressed in his letter to Stephan, Aug. 12, 1936, Reel 2.

converted paper locals in New York City, Burke began to employ Jewish and Italian organizers recruited from these unions, to organize bag, box, and novelty workers throughout the eastern part of the country.[26]

Of all the new ethnic stocks recruited into the union in the thirties, Negroes were probably the most alien and perplexing to Brotherhood representatives. Prior to 1933, the organization had had virtually no black members, but with its growth in urban areas, and especially with its emergence in the South after 1936, the situation changed drastically. On occasion union representatives encountered problems in organizing urban Negro workers. Still, most of the Brotherhood's contact with blacks occurred in the southern pulp, bag, and kraft industries, which grew enormously in the 1930s and proved a fertile ground for Pulp, Sulphite, and Paper Mill Workers' organization.

By the time of the war thousands of Afro-American workers toiled in pulp and paper mills in the South. Most of them performed the more menial tasks, working in the woodyards, on cleanup crews, on loading docks, and on janitorial teams. The South became a prime target of the Brotherhood; the new mills of Southern Kraft (a subsidiary of International Paper), Rayonier, Union Bag, and Champion Paper presented attractive opportunities for expansion and for stabilization nationwide. In most cases, the union found that the new facilities in the South were operated by national employers who were generally sympathetic to the Brotherhood's aims. Such was particularly the case when CIO unions posed a threat, as they did in Alabama, Florida, and Mississippi.[27]

Black workers, however, presented problems. Local white unionists, who insisted on affording them subordinate status, usually required blacks to organize into "A" locals, attached to but inferior to the white locals. Thus, Mobile Local 337 represented white pulp and paper workers at the Southern Kraft facility in that city, while Local 337-A represented the black workers. In addition white unionists insisted on controlling these "A" locals, either by impos-

26. Zieger, "Oldtimers and Newcomers," 192–94.

27. Figures for the number of black workers and unionists are extrapolated from Auditor's Report, 1941, Reel 1. See also Northrup, *The Negro in the Paper Industry*, 32–34.

ing a white leadership or by virtue of highly manipulative "tute-lage" of the blacks by white officers.

Wage rates and working conditions for black workers, quite apart from the inferior employment usually offered them, were decidedly poorer than those of whites. In general, there was at least a 15 percent difference in wages for similar work. In addition, blacks often did not receive proper job grading at the hands of white supervisors, so that they earned rates even lower than those contracted for the job performed. On virtually every count, black workers in the southern pulp and paper industry suffered from the classic disabilities: lack of seniority; unfair classification; disdainful treatment by supervisors and union representatives; low wages; and general discrimination. "You know, Mr. Burke," wrote union member Willie B. Williams of Local 395-A at Rayonier's mill in Fernandina, Florida, "we colored people have been kicked around long enough. I have to pay as much as the white man for anything I buy. But working it's different." When the mill ran slack, white workers regardless of seniority superseded black workers; in such cases Negroes were laid off or, as in Wilson's case, were assigned to lower-paying jobs. "I have to work for less just because I have a different color, and my own brothers [(] allow me to call them that [)] is taking my jobs from me." [28]

Such conditions made the black workers excellent recruiting material for CIO unions. In 1940, CIO operatives in the South created an unofficial body called the Southern Kraft Workers Organizing Committee (SKWOC) and sought to organize pulp and paper workers in the Gulf Coast states, concentrating on black workers. The target was well chosen. Not only were blacks the victims of discrimination and objects of contempt; in addition, they were quite often strategically placed to disrupt the production process. Especially in the modern mills arising in the South, the critical bottlenecks occurred at the very beginning and end of the paper manufacturing process. Thus, black workers who performed the onerous tasks of wood handling, debarking, chipping, and paper loading were often important in ensuring the success of union organization.

28. Brotslaw, "Trade Unionism," 141–42; Northrup, *The Negro in the Paper Industry*, 26–45; Willie B. Wilson to Burke, c. May 5, 1940, Reel 3.

CIO unionists accused the two AFL unions of collaborating with Southern Kraft to force black workers into their organizations through rigged certification elections and under-the-table negotiations. They objected to the practice of relegating Negroes to segregated locals. Burke and his associates perceived that CIO inroads among blacks constituted a major threat to the union's control of the southern pulp and paper industry. "We must organize the Negroes in all of our mills in the south," Burke asserted. Given blacks' key locations in the mills, "we are just leaving a vulnerable point for the c.i.o. by neglecting to organize them."[29]

Unfortunately the international union's efforts to make concessions to black workers antagonized many white workers in the South from the very beginning. In January 1935 Brooks described his attempt to preserve a small local at a paper mill in Roanoke Rapids, North Carolina. Although it employed only 140 workers, the approximately 70 whites insisted on a separate charter for the remaining blacks. "To you and I," observed Brooks to Burke, "This is one hell of a condition but the question of color is about as strong . . . as it was at the close of the Civil War." Whenever the international union sought to organize, white workers—or at least activist elements in the local unions—bristled at the mention of racial equality or interracial organization.[30]

The Brotherhood's attitude toward these developments was hardly characterized by an aggressively integrationist stance. Neither Burke nor his chief lieutenants quarreled with the basic wage differentials prevalent between white and black workers, priding themselves rather on their ability to secure better conditions for workers of both races. Brooks, who spent more than a year organizing 2,000 pulp and converting workers in a Savannah mill, expressed his views of the whole racial situation in a series of reports throughout 1939-40. Although at times he sympathized with the Negroes, criticizing white unionists for insisting on separate locals and for treating black workers unfairly when there were slowdowns, he ultimately considered them a burden that whites had

---

29. SKWOC *News*, May 6 and May 15, 1940, in the file for special organizer Wilford Smith, Reel 3; Burke to Homer Humble, Local 229, Prichard, Ala., March 29, 1940, Reel 4.

30. Brooks to Burke, Jan. 8, 1935, Reel 1.

to bear. The new Union Bag mill in Savannah employed nearly 2,000 workers, 450 of them blacks who performed mostly the outdoor and clean-up work. Such jobs provided blacks with a grand opportunity to improve themselves industrially, Brooks believed, but most of the Negroes were content with drifting from day to day. The organizer observed, "I find that while the negro, generally, is always quite ready to reach out a hand and accept anything that does not require thinking or effort . . . , he shows no desire or ambition to get into the fight and help carry the load." Since blacks seemed to thank neither the government nor the union for raising wages and improving conditions, their white co-workers were, in Brooks's opinion, justified in feeling "disgusted with [them]." In private correspondence with Burke, the veteran organizer stated matters even more crudely: "To hell with brother smoke. As a whole they are nearly hopeless."[31] Such open expressions of bigotry were unusual, however. While other organizers in the South, such as Wilford Smith, A.B. Hoff, and James Malin, shared the prevailing stereotypes about the role of blacks in the industrial labor force and did not question the propriety of discrimination and segregation, neither did they openly express contempt for Negro workers.

Burke himself never indulged in racial epithets. Throughout the organization of the South, he deferred to the sensibilities of white unionists, although he occasionally deplored the racism and bigotry that blacks encountered. When Negroes at one time protested unfair treatment by their white co-workers, he observed, "We shall have to do some educational work among our white members to get them to change their attitude toward their colored Brothers." He also took pains to explain to his socialist friends that the policy of discrimination and segregation did not emanate from the international union but rather was imposed on it by the southern locals. Still, his major concern remained the organization of an entire section of the country, and he did not permit qualms about the subsidiary issue of race to distract the union from coming to terms with major employers and from defeating the cio. He occasionally suggested more equitable treatment of Negroes; he occa-

31. Brooks to James Healy (copy), July 5, 1940; Brooks to Burke, Aug. 24, 1940, Reel 1.

sionally even warned white unionists that NLRB regulations, as well as common decency, forbade certain egregious forms of discrimination. But in the end he was always eager to reassure southern whites: "Of course, I fully understand that you boys down South understand this colored problem much better than I do."[32]

The alternative to acceding to the wishes of southern whites on racial matters was to attempt to rally whites and blacks together, as the CIO sought to do, in common hostility to the paper companies. Because Southern Kraft was willing to come to terms with the Paper Makers and with the Pulp, Sulphite, and Paper Mill Workers, this was not an option that Burke felt he could legitimately consider. The South had been the burial ground for many of organized labor's hopes over the years. No question was more incendiary or more fraught with peril than the race question. An aggressive stance on behalf of racial equality would have required years of confrontation, not only in the mills themselves, but in the factory cities and union halls as well. Burke believed that the favorable political and social climate of the 1930s would not last long. The union had to organize along the path of least resistance while it was possible to do so. If black workers joined the union, supported it faithfully, and heeded their international representatives, they would achieve improved conditions, greater opportunities, and industrial stability. For its part the Brotherhood would attempt to organize the South by pursuing maximum gain with minimum confrontation. Nevertheless, by late 1941, the union had organized only a small fraction of the blacks working in southern pulp and paper.

Whether the workers they encountered were blacks, women, Jews, or members of some other "exotic" group, organizers felt that theirs was a difficult and little-appreciated job. At the same time, Burke always had more applicants for positions than he could accommodate, for many workers saw a job with the union as their best opportunity to escape from the mills and to find more prestigious and more lucrative work. He warned prospective organizers to expect criticism and misunderstanding on the part of workers. Even so, as he noted to one of his vice-presidents in 1937,

32. Burke to Willie Williams, May 10, 1940, Reel 3 (first quotation); Burke to Ross Rabien, Local 272, Jarratt, Va., Jan. 29, 1940, Reel 4 (second quotation); also Burke to William Bohn, Aug. 8, 1938, Reel 7.

"Once you mention [that an organizing job is open] I am deluged with applications."[33]

At any given time in the late 1930s, the Brotherhood might have between twenty-five and thirty organizers. Of these, about a quarter would consist of local unionists on short-term leave of absence from their jobs, working for a few weeks to resolve local tensions or to help a visiting international representative with a strike or organizing campaign. Another three or four would typically be more veteran organizers who still held jobs in the paper mills. They would obtain longer leaves of absence, often periods of three or four months. These organizers traveled extensively within a given territory—in the Great Lakes region, for example—servicing locals, often without the supervision of a permanent representative. They faced conflicts between their usual jobs and their union work, for often they hoped for permanent positions on the union's staff and hence wanted to log as much time in the field as possible. But employers were not always willing to grant their requests, and family life suffered.[34]

The international officers, including auditor George C. Brooks, doubled as organizers and indeed spent most of their time on the road. As roving representatives, they covered vast amounts of territory. "We are battling away on a dozen fronts covering this vast North American continent," declared Burke in the summer of 1937.[35] Veterans such as Brooks and Vice-Presidents Herbert W. Sullivan, S.A. Stephens, and John Sherman were dispatched wherever trouble brewed or unusual opportunity beckoned.

In April 1941, for example, as the international union began its lengthy contract renewal sessions with more than 200 companies, the international officers were almost never at home. Vice-President William H. Burnell, who had responsibilities primarily in Canada, was out twenty-three days and covered more than 2,200 miles, holding thirty-two conferences and meetings that month. S.A. Stephens, whose work focused on Ontario and the Great Lakes states, held almost sixty conferences, spending the entire month on the road and logging more than 3,300 miles. Vice-

33. Burke to S.A. Stephens, March 6, 1937, Reel 1.
34. Rasmus Anderson's reports for 1937, Reel 1; Burke to Anderson, Aug. 2, 1937, Reel 1.
35. Burke to Dave Beck, July 8, 1937, Reel 2.

President John Sherman, whose responsibilities embraced the West Coast, which had about forty locals, found himself traveling more than 3,500 miles in the twenty-six days he spent on the road.

By September, the pace had mounted, with all vice-presidents spending almost the entire month in hotels, railroad cars, and conference rooms. Sherman put in more than 8,000 miles, while Stephens covered almost 5,500. Burke relied on the peripatetic Sullivan, whom he sent to virtually every locale to handle the knottiest disputes, the most contentious local unions, and most delicate organizing campaigns. None of the vice-presidents had an exclusive territory, however, and men who were expert in the New England pulp and newsprint field might find themselves negotiating contracts for set-up box workers or southern mill hands from one week to the next.[36]

The remaining corps of organizers—by 1940 usually ten or twelve—consisted of full-time employees of the international union on the payroll for an indefinite period. The usual route to a full-time organizing position involved service as an elected officer at a local union and subsequent temporary work with the international union. When he appointed organizers, Burke stressed that full-time jobs carried no job security and were subject to immediate layoff if business conditions turned downward. Still, the salary (about forty-five dollars a week, plus road expenses, for ordinary organizers in 1941), and the prestige and opportunities, were more than enough inducement for workers to grasp any opportunity to join the international staff.[37]

Most of the work performed by international officers and organizers was routine and undramatic. Since the international union's leadership avoided confrontation wherever possible, and since Burke prided himself on his "realistic" understanding of the economics of the pulp and paper industry, Brotherhood representatives were not expected to be proletarian heroes or charismatic orators. For every mass rally that had to be addressed,

36. Representatives' itineraries, transportation arrangements, hotel bills, and so forth are in their respective monthly reports, contained in the files under their names in Brotherhood Papers for each year. Those to which I have alluded in this paragraph appear in 1941, Reels 1 and 2.

37. Burke to George Gerrits, Sept. 30, 1936, Reel 3; Burke to O. Wing, April 20, 1940, Reel 5; George C. Brooks, International Auditor, Monthly Reports, 1941, Reel 1.

there were a hundred negotiating sessions to attend. In New England and in the Pacific Northwest, where by 1940 bargaining had become highly routinized and where little organizing work per se remained, international organizers served largely as contract administrators. In the more diverse Middle West, Vice-Presidents Raymond Richards and S.A. Stephens and organizers Valeria Brodzinski, Walter Trautmann, Keith Wentz, and others crisscrossed the area from Minnesota to Ohio, tending established locals in Wisconsin, Michigan, and Minnesota, fighting obdurate converted paper employers in Indiana and Illinois, attempting to organize Chicago's large box industry, ironing out jurisdictional problems with craft unions, keeping a wary eye on CIO activities, conferring with state and federal labor relations officials, attending local meetings, and generally exhorting, cajoling, and bullying workers to maintain their local organizations.[38]

Organizer Edward Mangan remained largely in the Middle West as well, specializing in the converting industry. Throughout the late 1930s he trekked thousands of miles in an endless series of one-day trips in a territory that stretched from eastern Pennsylvania to Illinois, organizing and servicing locals in Philadelphia, Allentown, Erie, Pittsburgh, the Cleveland-Akron area, Cincinnati and the Miami Valley, and small industrial cities in Indiana such as Shelbyville and Hartford City. In 1937 Mangan was almost perpetually on the road. In January he concluded agreements for a couple of multiplant Cleveland locals and settled a knotty

38. The correspondence and reports of each organizer and union official are grouped in a file under the person's name for each year. Thus it is possible to "follow" a particular organizer through the course of a calendar year by using the appropriate files of Brotherhood Papers. In later years organizers, some of whom eventually became international officers, retired or passed on. Each was honored or memorialized in the pages of the *Pulp and Paper Worker*. These little tributes supply details of the early days of organizers' activities for the union. See, e.g., *Paper Worker* 14 (Oct. 1959) on Walter Trautmann; *Pulp and Paper Worker* 15 (Aug. 1960) for Frank C. Barnes, Jr., a West Coast organizer in the 1930s who eventually became treasurer of the Brotherhood; *Pulp and Paper Worker* 17 (Dec. 1962), for William E. Riggs, a West Coast activist and official; *Pulp and Paper Worker* 21 (Dec. 1964) for a 30th anniversary sketch of Toledo Local 187, three of whose leaders eventually became international representatives; and *Pulp and Paper Worker* 23 (Oct. 1965) for a brief biography of Joseph Tonelli, a local leader and international representative in the 1930s who in 1965 succeeded Burke to the presidency of the Brotherhood.

grievance case in Rittman, Ohio. Throughout the spring he organized a local in Erie, Pennsylvania, and hammered out agreements involving small locals throughout Ohio and Indiana. In June, the Erie local experienced difficulties with its employer, who refused to bargain with the new union. Thus, in July, Mangan stayed with the workers and conducted a successful ten-day strike. The remainder of the year brought more of the same. In the fall, Mangan became ill, afflicted with perpetual drowsiness, probably the result of fatigue and of irregular eating and sleeping habits. He remained on the road, however, fighting his illness, eager to carry on the union's mission of providing sober, cautious, effective service not only to its members but also to the industry in which they toiled and to the public at large.[39]

Organizers and international officers regularly reported directly to Burke. By 1941, the union collected per capita taxes of about $40,000 monthly. It boasted more than 300 local unions in almost every state, in most cities, and in more than half the Canadian provinces plus Newfoundland. It employed in that year about thirty-five organizers. Yet there was still no regional or product suborganization within the chain of command. All files were kept in Fort Edward, and all requests for information came to Burke, to be handled directly by him or, in his absence from the office, by his secretary Francis Tierney or by his wife, Bessie Leon Burke. An enormous flow of correspondence, reports, and memoranda perpetually poured into the little office. An unusual crisis might necessitate a telegram, and in rare emergencies, an organizer might presume to make a long-distance phone call to Fort Edward. Still, the U.S. Mail carried most of the union's internal communication.

Representatives shared a common view of the goals and practices of organized labor. Veteran officers had attended the same school of hard knocks that had educated Burke, and they needed little encouragement from him to be patient in organizing and moderate in negotiations. Newer representatives fit into the same mold; Burke relied on his experienced cohorts to recommend local unionists who displayed the desired qualities for organizing

39. Mangan, Report, Brotherhood *Journal* 21 (Jan.-Feb. 1937), 10; Mangan file, Brotherhood Papers, 1937, Reel 1. By 1938, he had recovered sufficiently to be dispatched to Texas to help organize Local 305 in Pasadena. The trip to Texas, a big event for the Ohioan, was commemorated in the Brotherhood *Journal*.

posts. Temporary organizers, imbued with some of the fiery enthusiasm of the 1930s, sometimes grew impatient with the international union's sober high-mindedness. But temporary organizers and new representatives who harbored hopes for careers in the international union quickly learned to curb their more heroic notions.

Consequently, Brotherhood representatives did not perceive themselves as fiery champions of the working class. Rather they were professionals, men and women who understood the constraints upon business as well as the needs of workers. They prided themselves on their detailed knowledge of the industry. Implicitly they viewed themselves as "managers of discontent," people as responsible for calming excited workers as they were for gaining concessions from unyielding employers. Union representatives sought the best possible contracts for the workers they represented, of course. But as Burke reminded his staff, it was a mark of "skilled and experienced" organizers that they also be able to conduct, when necessary, the "retreats [which] are often but good strategy."[40]

Above all, organizers had to be good salesmen. Burke believed that a labor representative had to develop "a 'line' the same as an insurance agent . . . or some of these fellows on the road selling Dill's pickles." Workers had to be convinced that a financial investment in the union paid off; employers had to be persuaded that the union would not radicalize or disrupt and that it could actually lead to increased productivity and profits. In either case, the organizer, like the insurance agent, but unlike the man peddling pickles, purveyed the hardest commodity of all for the drummer to move—an intangible good that promised benefits in the future but required payment in the present. As organizer A.B. Hoff remarked, the organizer had a difficult task indeed—"to sell job insurance to the paper workers."[41]

Representatives thus often viewed the pulp and paper workers as clients or "union customers." But they were frequently irksome and irresponsible clients, often behaving irrationally. Workers too often acted on false information. They failed to grasp economic realities. They demanded endless service, yet they showed little

40. Typescript of Burke, Report to 19th Convention, Sept. 8, 1941, Reel 9.
41. Burke to Raymond Leon, Nov. 26, 1937, Reel 2; A.B. Hoff in *Journal* 21 (Sept.-Oct. 1937), 6.

gratitude and had to be coerced by employers into paying their dues. Representatives often described workers as children who expected organizers "to run with a nursing bottle every time some member has a belly-ache." "An organizer," Burke advised one of his discouraged representatives, "must be something of a psychologist. . . . Kid the people along at times; coax them at other times; and drive them at various times."[42]

Brotherhood representatives of course believed in the mission of the labor movement. Their belief, however, was primarily in the *union* as a concept and as an institution rather than in the workers as a collectivity. While workers had at times failed the union, the union had never failed the workers. Workers might become discouraged or apathetic. They might give in to an employer's threats or blandishments; they might also fail to pay dues, cease to attend meetings, or forget their oath of loyalty to their fellow workers. Each organizer, no matter how new in the field, had his stories of workers abandoning the union, forsaking it for reasons of momentary advantage or from discouragement. Even under the best of circumstance, Burke believed, in union activities "it is only a very small minority of people who are willing to work constructively toward the attainment of an objective."[43]

If workers were often apathetic and lackadaisical in fulfilling their union obligations, they also frequently indulged in rash and ill-advised activism. They felt that the very fact of organization entitled them to instant recognition and to improvements in wages and conditions. Negotiations between international representatives and employers that were conducted with good cheer seemed to workers to smack of collusion. Sometimes they were eager to go out on strike, disregarding constitutional procedures and expecting massive support from the international union. In short, workers often lacked the discipline, foresight, patience, and commitment needed to provide a solid footing for local unions.

Still, of course, workers deserved organization. Burke fancied himself an old-fashioned democrat, pointing to New-England-style town meetings and the yearly elections in Fort Edward as models of grass-roots democracy that workers might well emulate. Again

42. Morris Wray to Burke, July 2, 1938, Reel 3; Burke to Stephan, July 29, 1936, Reel 1; Burke to Leon, Aug. 23, 1937, Reel 2.
43. Burke to William Hineline, Sept. 29, 1939, Reel 2.

and again he chastised local unionists who sought financial or organizational help from the international office, warning them that they must learn to run their own affairs and to take responsibility for their own decisions. He grew angry and frustrated at tales of local unions atrophying after an initial rush of enthusiasm. In 1935, he advised organizer Walter Trautmann that "the great problem in trade union organizing work is how to maintain interest among the members from month to month." After an exciting period of organizing, with perhaps a strike and an initial contract spurring interest, "the average human being is inclined to become indifferent and lose interest." Informed of the impending collapse of a large Ohio local, Burke observed that once again "it is the old story of the workers being to blame. They join the union . . . and then, after they get everything they can . . . they drop out." The organizer's job was difficult and nerve racking. He or she had to cajole the worker, to impress the employer, and to cope with a skeptical public. Too often it seemed that the people most benefiting from his efforts were the least understanding and the least appreciative.[44]

Brotherhood representatives had plenty of opportunity to nurture their feelings; the men who traveled for the union spent a great deal of time alone. Perpetually strangers, frequently in tense situations, they usually found themselves without companionship. Even fellow unionists could not always provide easy camaraderie, for perpetual jurisdictional problems with the craft unions and the Teamsters created personal frictions, while the city centrals in the larger towns remained closed circles of cronies, providing little hospitality to any but the most grizzled veterans on the Brotherhood staff. Then, too, the organization found itself in the 1930s entering cities in which it had never before maintained organizations.

Prior to the great expansion after 1933, the small number of representatives that the union kept on the road could usually find old comrades, sympathetic fellow unionists, or former co-workers in the relatively constricted geographical region that contained the pulp and paper mills. But the golden opportunities of the thirties sent union organizers to many strange and exotic places under circumstances that encouraged their feelings of apartness and

---

44. The first quotation appears in Burke to Walter Trautmann, June 3, 1935, Reel 3; the second is in Burke to Maurice La Belle, July 20, 1935, Reel 1.

solitude. Thus 1939–40 found international auditor George Brooks living in a hotel for more than a year in Savannah, Georgia, attempting to organize the 2,000 workers at Union Bag and Paper's big new mill in that city. During the year, Brooks suffered a recurrence of a leg injury. Family problems forced him to bring his wife down from New Hampshire, and she later underwent a complicated appendectomy in the southern city. Brooks was an acerbic, opinionated man, but perhaps his isolation and the inherent tension of attempting to organize inexperienced workers, black and white, while dealing with a difficult and uncompromising employer darkened his view of his charges and added pungency to his assessment of the capacities of southern workers to bear the burdens of unionism.[45]

Organizers often expressed bitterness and dismay about the workers they sought to organize and represent. New Yorker Morris Wray, an outspoken leftist and a firm believer in the union's mission, suffered many setbacks in New Jersey in 1938 and became depressed. Noting that box workers in New York and Philadelphia had criticized the international union for failing to organize the New Jersey shops, Wray reported that he spent several days canvassing in North Jersey, visiting plants, talking to workers, going from house to house after working hours. For his pains, he reported, "I've been driven away and chased." In south Jersey he found that "anyone that talks now to people about joining a union is looked upon as crazy." Frustrated and disillusioned, Wray was ready to throw in the sponge. "I don't know how to work this business any more."[46]

The organizers unburdened themselves to President Burke, expressing in their reports and letters both their self-doubt and their self-righteousness. Anger and frustration they vented primarily at the local workers, who so often seemed unreasonable, indifferent, and demanding and who found succor and camaraderie among family and friends, while the organizer ate a lonely meal and trudged back to his hotel to fill out a report. Sympathetic to these stresses, Burke sought constantly to protect his organizers

45. Brooks's tribulations are followed in Brotherhood Papers, 1939 and 1940, Reel 1, file for Brooks.
46. Morris Wray to Burke, Aug. 6 (first quotation) and July 2, 1938 (second quotation), Reel 3.

from the demands of local unionists. "Our great little organizer, Jacob Stephan, one of the finest labor organizers that has ever lived," he informed a local in Escanaba, Michigan, "has cracked under the terrific strain. . . . and today he is but a shell of his former self."[47]

Despite the international union's cautious approach to its tasks, few labor organizers in the 1930s could escape confrontation entirely. Company goons threatened Raymond Richards repeatedly as he sought to organize the mills of central Wisconsin, while organizers on the West Coast found arrogant Teamsters threatening to support their jurisdictional claims in California with violence. "Just came out from the jug for giving out circulars in Delair, New Jersey," reported Wray in April 1938. "They had us finger printed and everything that goes with arresting criminals."[48]

More dangerous, though, than bullyboys and policemen were the hazards of travel. With the union's expansion, organizers were increasingly forced to rely on private automobiles and even on air transportation. The severe weather of the union's northern territories often made driving hazardous. Sullivan lost several teeth and a good deal of blood in a serious accident in Washington State in 1937, while John Sherman, who found rail travel simply inadequate in making the rounds of his scattered Pacific Northwest locals, had several auto accidents that varied in seriousness. On a narrow Michigan highway Walter Trautmann skidded and collided with another car, suffering cuts and bruises. James Killen, a West Coast organizer, found himself hospitalized after a bad mishap. The same accident that injured Sullivan killed Pacific Northwest organizer Dave Beck. The international union provided no insurance for its employees, but Burke doled out hospital and funeral expenses to the sick, the injured, and the survivors.[49]

Burke himself regarded travel by air as a particular trial. Colleagues warned him of its dangers. He believed that every time he flew to the Coast and back safely the odds against an equally safe return the next year diminished. "I am becoming just a little afraid of these airplanes," the president-secretary admitted in

47. Burke to Einar Beck, Jan. 21, 1936, Reel 2.
48. Morris Wray to Burke, April 11, 1938, Reel 3.
49. Burke to James Killen, Dec. 14, 1937, Reel 2; Walter Trautmann to Burke, Aug. 2, 1941, Reel 3.

1937. "There have been so many accidents." But train travel was too slow, and the Pacific locals and employers expected an annual visit from the international president. The price of success and growth for Burke as well as for his corps of representatives was discomfort, tension, and sometimes even fear.[50]

Throughout the late 1930s and into the early 1940s, Burke and his representatives plied their trade. Missionaries and salesmen, they canvassed the country, building the union in their fashion. From Newfoundland to Los Angeles, from British Columbia to Florida, they dealt with one of the most diverse jurisdictions in the American labor movement. Black and white, Jew and Italian, beaterman and cardboard stacker, workers of all races and backgrounds found their way into an organization still led by crusty veterans of the teens and twenties. Fort Edward remained at the center of their world, even as they pursued pulp and paper workers far from the industry's declining northeastern sector.

50. Burke to Raymond Richards, May 28, 1937, Reel 1; I have quoted Burke to James Killen, Oct. 30, 1937, Reel 2. "In those days," remarks novelist Anthony Powell, describing a trip to California in 1937, "flying from coast to coast was regarded, with its risk of hitting the Rockies or Alleghenies, as quite an adventure and entailed three descents for refueling" (*The Memoirs of Anthony Powell*, vol. III: *Faces in My Time* [New York: Rinehart and Winston, 1980], 69).

6

Moving Ahead on
All Fronts

I N 1937 THE BROTHERHOOD'S STRATEGY BEGAN TO PAY SUBSTAN-
tial dividends. By the fall of that year, membership more than
doubled, reaching the 30,000 mark. The union revived its
older locals in the Northeast and continued its impressive gains
in Wisconsin. It also secured union shop agreements in the Pacific Northwest, fastening its hold on this important production
area. In addition, in the late 1930s it made its first successful forays into the South, the section of greatest promise for the expansion of the paper industry, and devised methods to consolidate,
rationalize, and expand its increasingly important activities in the
urban converted paper field.

In making these gains, the international union reaped the benefits of the turbulent labor scene of the late 1930s. The threat of
CIO activity impelled both employers and the AFL to encourage
the paper unions to expand in the paper industry. With the increasingly effective protection of the NLRB and the example of fellow workers in other industries challenging employers and making contract gains, pulp and paper workers proved receptive to
the unions' appeal. Burke and his representatives avoided the passionate militancy so prevalent in the late 1930s, benefiting from
the general milieu of unionist fervor but relying, as usual, on caution and patience rather than on fiery appeals. The organization

came increasingly to depend on union shop contracts that international representatives negotiated with multiplant employers or
with employers' associations to capture pulp and paper workers.
While membership gains were unmistakable, reliance on these
methods permitted critics to attack the union for its inability to
tap workers' enthusiasm and for the inevitably cautious posture
in bargaining and representation that employer-negotiatied union
shop agreements required.

In Burke's ideal union, workers would pay their dues without
coercion. In the real world, he believed, the only way that the
international union could achieve solid financial footing was
through union shop agreements with employers. Theoretically,
some sort of checkoff system by which the employer deducted and
turned over to the union the worker's monthly dues would have
been best, but in the 1930s Brotherhood rarely achieved the checkoff. More common were arrangements in which the union secretary or treasurer collected the dues and gave the worker a receipt
in the form of a stamp, to be pasted in his dues book. If a worker
fell behind and if he or she ignored warnings, the employer, under the terms of the union shop agreement, would discharge the
offender.[1] Such agreements were standard in New England, New
York State, Wisconsin, the Pacific Northwest and, increasingly by
1941, in the large southern mills. Writing to Wisconsin organizer
Raymond Richards in 1936, Burke paid homage to the Hoberg
Paper Company of Green Bay, which had just renewed its long-
standing union shop contract with the organization. "Without the
cooperation of the company in keeping the workers organized,
it would be a constant struggle even to keep a local alive there,"
he observed. Furthermore, the union shop agreements in the older
mills provided the resources that enabled the Brotherhood to extend organization into other sectors.[2]

At the same time, resort to union security agreements distressed
the Brotherhood veterans. It was a shame, Burke believed, that
workers had to be coerced into the union. "If the workers could
only organize without [the union shop] agreements and remain

1. Typical union shop agreements: Burke to D.W. Crawford, Dec. 23, 1937, Reel
7; Burke to George Brooks, July 12, 1937, Reel 1; I.H. Copeland of Canadian International Paper to F.P. Silver, June 2, 1937, Reel 7.
2. Burke to Raymond Richards, Feb. 17, 1936, Reel 1 (quotation); Burke to S.A.
Stephens, Nov. 19, 1937, Reel 1.

organized," he declared, the unions would not have to discipline
their members or to accept unpopular contract provisions. "However, just as long as American workers have to have the employers keep them in the union, then we must live up to our agreements."[3] In addition to his basic misgivings about the union shop,
Burke regarded it as a divisive and even potentially destructive
issue in newly established locals. The union shop, he continually reminded new members of the union, had been achieved in
the older sections only after years of struggle, organizational maintenance, and repeated proof to employers both of the union's per-
severance and of its usefulness. It was hardly realistic for a newly
organized group to demand the union shop. True, workers insisted
that representatives bargain for union security. Again and again
local unionists complained that in the absence of the union shop,
workers refused to pay dues, fearing that their free-riding fellow
workers would consider them foolish to pay for something that
others received free of charge. Local unionist Nelson Moon of Local
240 in Connellsville, Pennsylvania, reported, "Several have said
that they wasn't paying . . . dues till the union shop agreement
was signed. Some others say, 'What is the use of paying dues when
they can have as much say so as Union members can in a shop
where you are not forced to become a Union member.'"[4] These
arguments never failed to annoy Burke. "If I had my way," he told
Wray in 1939, "I would have it understood with every newly
organized local that we would not even approach the company"
concerning union security until the union had been maintained
in good standing for at least six months. The union shop, he told
a Cincinnati unionist, came as the *result* of vigorous organizing
and faithful maintenance of a local union. Once these conditions
prevailed, "we can say to the company that the overwhelming majority of the workers in the shop are members of the union in
good standing" and can legitimately bargain for formalizing the
union shop situation.[5]

Burke feared reliance on the union shop for still another reason. Compulsory membership brought workers into the union

3. Burke to Morris Wray, July 24, 1939, Reel 3.
4. Nelson Moon to Burke, c. Oct. 12, 1936, Reel 2.
5. Burke to Morris Wray, March 10, 1939, Reel 3; to Joseph Weingartner, Local 206, Dec. 22, 1936, Reel 2.

who felt no loyalty toward it. If anything, some workers resented being forced to pay dues in order to keep their jobs. Many of them became apathetic. Others, however, sought to exploit those who were disaffected and kept locals in constant turmoil."The compulsory membership clause forces a large number of workers, who know nothing about the union and who care less, to become members," Burke observed. Extremists "use these union conscripts to put over their schemes and programs" and to abuse the international union.[6]

Still, the union depended on the union shop to maintain itself. The issue might be divisive in new locals. It might provide tinder for extremists. It might severely limit the workers' ability to confront their employers. Indeed, it might (a possibility to which Burke did not address himself), and often did, modify to the point of collaborationism the international union's bargaining stance vis-à-vis employers who loyally enforced the union shop. Despite these drawbacks, as far as Burke was concerned, "we have to fight to get a union shop clause . . . so that we can get the help of the company in maintaining unions for us."[7]

Thus although Burke both believed and hoped that trade unionism was a democratic enterprise, he felt it unrealistic and even dangerous to expect too much of workers' self-motivation and capacity for self-government. Whatever his convictions about mass organization, about the eventual goals of the socialist movement, and about the sanctity of labor, he always affirmed that the Brotherhood, as an AFL affiliate, was "committed to a policy of signed written union shop agreements . . . and to a policy of abiding by these agreements, both in spirit and in letter."[8]

Burke believed in a democracy of the committed. The stalwarts who had manned the picket lines, who had upheld the union in its times of trial, and who carried the union banner, even in retreat, were the real union men and women. Hence he saw nothing wrong with taking stands that would limit the formal participation of union members in decision making. He urged international

6. Burke to Frank C. Barnes, Jr., April 23, 1937, Reel 1.
7. Burke to Nils Gunderson, Local 42, Eau Claire, Wisc., May 29, 1939, Reel 4.
8. Burke to W.B. Gorbutt, Oct. 30, 1937, in Burke to all international representatives, Nov. 5, 1937, Reel 3.

representatives to promote their selection by local unions as delegates to the biennial convention; by this means the current leaders of the international union would be better able to oppose efforts to reduce dues. Multiplant agreements negotiated jointly by
local union representatives and international officials should be
binding, he maintained, without subsequent submission to the
local unions for referendum votes. He kept control of the organizing force, ensuring that appointees would be loyal to his administration.

Despite his ideals of union democracy, by the 1940s Burke had
concluded—however reluctantly—that workers often simply did
not know what was in their best interests. Some unions, he noted,
had become "virtual dictatorships." No one liked dictatorships,
but "they had to become that in order to survive." Although the
Pulp, Sulphite, and Paper Mill Workers represented no dictator-
ship under John P. Burke's leadership, the union was far from the
model of democracy he would have liked. Critics might complain
that the Brotherhood too closely adhered to its AFL traditions,
that it grew too close to employers, and that it relied too heavily
on union security devices. Burke's answer was simply that when
unions grew bureaucratic and oppressive, they did so not because
evil men sought power but because "the rank and file of the members act so foolishly that somebody has had to step to the front
and save them from themselves." Labor had sacrificed too many
martyrs to permit its proudest embodiments—the trade unions—to
fall prey to shortsighted and irresponsible members. If it was sometimes necessary to save the union from the workers, Burke was
entirely willing to undertake the task.[9]

While the pulp and paper workers themselves sometimes seemed
less than enthusiastic about enlisting in and supporting their locals, employers in the industry took up some of the slack. Throughout the 1930s, the union sought to persuade pulp and paper manufacturers of the benefits of trade unionism. Certainly, Burke and
his leading associates stressed the union's role as a provider of stability and discipline in the mills. Repeatedly they hammered home
the lesson that union workers took pride in their tasks and wanted
to give full service for their wages. "Union men," advised Burke,

9. Burke to Wray, Jan. 31, 1941, Reel 3.

"should endeavor to give the best possible service, because the better workmen our union members are, the better it is for the Union."[10]

Union officials placed great stock in influencing company officials to think well of them and their organizations. "We believe in working in cooperation with the employers," Burke declared. During the NRA period he, Sullivan, Matthew Burns, and other paper union representatives stressed their moderate goals and cautious methods at industry code hearings, taking pains to contrast these qualities with the irresponsible brand of unionism surfacing throughout the country in 1933 and 1934.[11]

Employers reciprocated. Apprehensive about the turbulence that the NRA had triggered and, later, about the rise of the aggressive CIO, paper industry employers lauded the Brotherhood for its sanity and responsible attitude. The Pulp, Sulphite, and Paper Mill Workers increasingly seemed a palatable antidote to the laborite radicalism that afflicted the country and threatened to spread into the pulp and paper industry. Throughout the mid-1930s, Burke enjoyed an extremely cordial and mutually supportive relationship with key executives in the Canadian and American paper industry. Thus a manufacturer in the state of Washington declared in 1937, "I believe it is vitally important that *we* keep the Longshoremen's Union from horning in on the pulp and paper mill set up, and that we do not relinquish any jurisdiction to the Longshoremen."[12]

Nor were the difficulties associated with rival unions the only source of employers' gratitude to Burke and his union; the Brotherhood could also moderate the impatience and restiveness of a company's employees. "When the rank and file of union members catch [sic] your common sense philosophy, all resistance to unionism [on the part of employers] will disappear," predicted the person-

10. Burke to Douglas Grant, Local 351, East Braintree, Mass., Nov. 11, 1939 (quoted); Burke to H.I. Houlette, Vice-President of Ohio Boxboard Company, March 4, 1938, Reel 7.

11. Burke to Eleanor Laubach, Local 310, Middletown, Ohio, Feb. 13, 1941, Reel 5. Burke, Statement, Hearing 405-1-04 (Paper and Pulp Industry—Proposed Revisions), Sept. 14, 1933, NRA Records, Records Maintained by Library Unit, Transcripts of Hearings.

12. Robert B. Wolf, Manager, Pulp Division, Weyerhaeuser, Longview, Wash., to Burke, March 1, 1937, Reel 7; emphasis added.

nel manager of a Wisconsin mill. And the president-secretary was in complete agreement, urging companies to cease opposing the requirement that foremen and supervisors join the union. Far from wanting these management representatives *out* of the union, companies should welcome their membership, for they would "act as a check and a balance at union meetings for those less experienced men who may desire to take extreme action at times."[13]

This mutually supportive relationship between employers and the union manifested itself throughout the post-1933 period, although it became increasingly evident in the later years. The Brotherhood's reputation for moderation and realism stood it in good stead in its efforts to reestablish itself in the old northeastern pulp and paper region. The union gained early success in reviving some of its moribund locals in New England and in New York State, renewing union shop agreements that provided regular financing for expansion elsewhere. At the heart of this process was the organization's continuance and extension of agreements with Canadian companies, whose workers composed almost 25 percent of the union's total membership and per capita tax receipts. The large Great Northern newsprint mills at Millinocket, East Millinocket, and Madison, Maine, which employed a total of 1,500 workers, also contributed a steady source of income.

Revival of other locals in the old pulp and paper towns of the Northeast, however, was more problematic. International Paper had closed sixteen of its mills in this area during the great strike of the 1920s, and its remaining plants in the United States, at Fort Edward, Palmer, and Niagara Falls, New York, and at Livermore Falls, Maine, produced for the company's troubled book and bond division, an unprofitable operation throughout the 1930s.[14] Eventually, in 1937, International Paper did sign union security agreements with the unions for these plants and for its large Canadian newsprint operations, but only after the union had made substantial headway throughout the pulp and paper field.

13. W.F. Ashe to Burke, Nov. 29, 1937, Reel 7 (first quotation); Burke to George Bearce, manager, Maine Seaboard Paper Co., Bucksport, Me., Dec. 23, 1937, Reel 7 (second quotation). Most of Burke's correspondence with company officials and trade association executives is found on the final reel for any given year under the heading "Companies."

14. Burke to R.J. Dickson, Jan. 21, 1937, Reel 7 (quoted); "International Paper and Power," *Fortune* 16 (Dec. 1937), 131–36, 226, 229, and 230–32.

Another eastern giant, the St. Regis Paper Company, also agreed to union contracts in 1937 for its remaining northern paper mills, as did the Oxford Paper Company. These agreements anchored the union's northern flank, assured the constant flow of per capita tax from about 4,000 workers, and provided powerful arguments for union representatives in their dealings with other large concerns.[15] Still, aside from Canadian newsprint operations, the Northeast was a declining pulp and paper sector, and the organization of the lake states, the Pacific Northwest, and later the South was essential if the union was to reassert itself as a major force in the industry.

Of these important territories, Wisconsin, Minnesota, and Michigan proved the most difficult. But in 1933 and 1934 Jacob Stephan succeeded in reviving a number of locals that remained functioning after the end of the NRA. Thereafter the Brotherhood won union shop rights at Mosinee, Wisconsin, through an NLRB election and in 1935 and 1936 entered the hitherto antilabor Marathon Mills in Rothschild and Wausau. The battle to win rights at Nekoosa-Port Edwards took longer. NLRB certification did not come until 1938, but union mills at Wisconsin Rapids, Green Bay, and International Falls and in several Upper Peninsula locations solidified gains in the Great Lakes area. The threat of CIO incursions in 1937 spurred organization throughout the Fox River Valley and added at least six large mills to the union's purview. A number of area employers were more than willing to enter into union shop agreements with the Pulp, Sulphite, and Paper Mill Workers and with its sister union, the Paper Makers. "Union agreements are being signed left and right," reported AFL functionary Ervin Wheelock of Neenah-Menasha, which was in the heart of northeastern Wisconsin's paper industry. "It's all very encouraging," he added.[16]

Adding to success in these older areas of strength was the union's highly effective penetration of the pulp and primary paper indus-

15. Agreements with St. Regis, Great Northern, International Paper, and Oxford Paper are all in Brotherhood Papers, 1937, Reel 7; also Burke to George Brooks, July 12, 1937, Reel 1. The membership figures are extrapolated from Auditor's Reports, 1938 and 1941, Reel 1.

16. Graham, *Paper Rebellion*, 12–14; Raymond Richards, typescript, Report to 18th Convention, March, 1937, Reel 7; Ervin Wheelock to Burke, March 29, June 16, and Aug. 14 (quoted), 1937, Reel 4.

try of the Pacific Northwest. By the end of 1937, the old north-eastern, Great Lakes, and Canadian pulp, paper, and newsprint locals accounted for about 60 percent of the total per capita tax collected by the international union. In that same year, the locals in Washington and Oregon contributed an additional 15 percent. This development was remarkable since, in contrast to the organization in the Northeast, the international union had no membership, no local unions, and no formal presence of any kind on the coast at the time of the passage of the NIRA in 1933.[17]

The pulp and paper industry in Washington and Oregon was unique for several reasons. Its mills were of relatively recent vintage, and it produced largely for the West Coast regional market and hence was relatively impervious to national competition. The management of the twenty to thirty mills that produced the bulk of the region's pulp and converting paper tended to be progressive, innovative in labor policies, and uninterested in turning the paper industry into a battleground between labor and management. The Crown-Zellerbach interests, which accounted for about 60 percent of the region's production, the Rayonier Company, and the pulp division of the Weyerhaeuser Timber Company, had little intrinsic hostility toward organized labor. They shared a keen awareness that low labor expense prevailing in their modern and highly mechanized mills would permit relatively flexible personnel policies and receptivity toward the unions. They all wanted to avoid the industrial bloodletting that had characterized the industry in the East in the teens and twenties and that afflicted the West Coast timber, lumbering, longshoring, and maritime industries in the 1930s.[18]

Of particular importance in easing the path of the paper unions on the coast was the presence of Robert B. Wolf, the manager of Weyerhaeuser's pulp division in Longview, Washington. Wolf had been trained as an engineer and had been associated with the engineering movement in the World War I era. He had collaborated

17. Burke to Green, June 27, to James A. Taylor, June 29, 1933, Reel 3; membership figures are culled from the monthly reports of international auditor George C. Brooks, which appear on Reel 1 for each year.

18. Graham, *Paper Rebellion*, 31–36; Kerr and Randall, *Collective Bargaining in the Pacific Coast*, 1–5; Guthrie, *Economics of Pulp and Paper*, 6–11. The atmosphere of labor unrest in the region in the 1930s is described in Christie, *Empire in Wood*, 152–214 ff.

with Herbert Hoover in efforts during the early 1920s to secure support from both labor and management for new patterns of industrial relations that were to be based on increased productivity, worker representation, and application of technical expertise. As manager of the Spanish River Paper Company in Ontario, Wolf had had pleasant relationships with the paper unions and in particular with Vice-President Sullivan. He viewed the paper unions, and especially the semiindustrial Pulp, Sulphite, and Paper Mill Workers, as ideal vehicles for harmonious industrial relations on the coast. His previous experience with the union had demonstrated that it faithfully adhered to its written contracts, was willing to understand the employer's needs, and was politically and socially conservative.

At the same time, the Brotherhood could function on a plant-wide and even an industrywide basis, free from the entanglements of the diverse craft unions that made the AFL so often unattractive to mass production employers. Indeed, from Wolf's point of view, the Pulp, Sulphite, and Paper Mill Workers represented precisely the blend of vertical organization, responsibility, and conservatism necessary to make the pulp and paper industry of the Pacific coast a model of industrial relations. Thus, as workers stimulated by the NRA and by the organizing fervor prevalent on the coast in 1933 flocked into local unions hastily called into being by local activists and by AFL functionaries, Wolf spoke frequently with Burke on the telephone, assessing the local situation, encouraging the unionist to commit his organizations's resources to a major effort in the Northwest, and pledging his discreet support in overcoming the prejudices of fellow employers. "If you succeed in unionizing the entire pulp and paper industry . . . and can demonstrate through the industry how cooperative work can really be successfully accomplished through the unions you will render a real service . . . ," he said encouragingly. "More power to you."[19]

Wolf's enthusiasm found its target. By the end of 1933, with the help of AFL representatives and newly appointed temporary organizers, the Brotherhood had established local unions at Longview, Port Angeles, Hoquiam, and Shelton in Washington and in Oregon City, Oregon. Through 1934 it pressed into other pulp

19. Burke to Green, June 27, 1933, mentioning phone calls; brief notes of phone messages, Sept. 7–8, 1933, Reel 3; Wolf to Burke, Sept. 18, 1934, Reel 6.

and paper communities. At the same time, the NRA and the turbulent labor situation on the West Coast, which culminated in the violent and divisive strike of the International Longshoremen's Association against coast shippers and terminals, stimulated union interest among workers throughout the region. Although the pulp and paper mills were relatively new large-scale employers on the coast, the mill towns were closely connected geographically and by tradition to the logging, lumber, and sawmill industries, with their heritage of labor conflict, as well as to the turbulent longshoring and maritime trades. Although several pulp and paper employers sought to channel their restive workers into newly created company unions, the sentiment in favor of bona fide organization was too widespread and the supporting laborite milieu of the Northwest coast too pervasive for these initiatives to succeed. An activist in Everett, Washington, observed to Burke in the spring of 1934, "Our local looks for a rapid increase in membership in the next month or two, as interest in unionism here is getting more intense every day."[20]

From the outset, the pulp and paper workers of the West Coast insisted upon mutual cohesiveness. They were determined to resist the inroads of craft unions and to organize on an industrial basis, regardless of official AFL policy. At the same time, the workers in the various mills of the Northwest early joined together in a regional bargaining arrangement. In October 1933, under the leadership of Ben T. Osborne, executive secretary of the Oregon State Federation of Labor, and unionists in Salem, Oregon, fifty-four representatives of ten Northwest locals gathered at Longview, Washington, and formed the Pacific Northwest Conference of Pulp and Paper Employees, a regional group within the AFL unions. Throughout 1933 and 1934, the international unions, the Pacific Northwest Conference, and the region's employers moved toward the creation of a formalized regional bargaining structure.

For the employers, led by the Crown-Willamette and Crown-Zellerbach interests, a regional arrangement was attractive for several reasons. Not only were these international unions free of the aggressive activism that shook the sawmills and docks of the West Coast, they also possessed a broad jurisdiction that could free the

20. Kerr and Randall, *Collective Bargaining in the Pacific Coast*, 1–5; Graham, *Paper Rebellion*, 31–33; Lyle Winlesky to Burke, April 21, 1934, Reel 3.

industry of the incessant wrangling and complex local bargaining that so often soured employers on AFL organizing activities.
In the fall of 1933, Alexander Heron and J.D. Zellerbach met Burke
at code hearings in Washington and found his grasp of the dynamics of the industry, his self-proclaimed realistic understanding of employer's problems, and his castigation of extremists in
the labor movement reassuring. After a round of presentations
involving hundreds of paper company executives in September
1933, Burke reported, "If we use our heads now and do not make
wild demands, I feel that we have an opportunity to organize the
whole industry."[21]

The integrated approach to bargaining culminated in early August 1934 when delegates from nineteen locals of the Paper Makers and the Pulp, Sulphite, and Paper Mill Workers met with a
committee of company executives led by Zellerbach of the Crown
empire. With Burke heading the union's bargaining team, the two
sides spent three days hammering out the Uniform Labor Agreement. This document was binding on all the local unions and
all of the employers belonging to the Pacific Coast Pulp and Paper Manufacturers' Association, an organization that soon enrolled all but a couple of small mills in Washington and Oregon.
The agreement recognized the paper unions as bargaining agents,
granted wage increases, provided for three paid holidays a year,
and created a joint committee to set wages for various jobs in the
mills. It also created a grievance system that included a Joint Relations Board.[22]

Throughout the 1930s, the Uniform Labor Agreement, annually modified and renewed in elaborate contract negotiations usually held in Portland, remained the cornerstone of labor relations
in the pulp and paper industry on the West Coast. By 1941, as
many as thirty-four plants fell within its provisions, each with locals of the two international unions. In the 1936 agreement the
unions secured a maintenance-of-membership clause and found
the companies vigorous in policing this key provision. The more

21. Telephone interview with George W. Brooks, Professor, New York State
School of Industrial and Labor Relations, Cornell University, Feb. 7, 1979;
Graham, *Paper Rebellion*, 31–34; Burke's quoted statement appears on p. 33.
22. Graham, *Paper Rebellion*, 33–34. Graham lists 13 companies involved, several of which, notably Crown-Zellerbach and Rayonier, possessed several mills
that came under the terms of the Uniform Labor Agreement.

than fifteen locals of the Pulp, Sulphite, and Paper Mill Workers in Washington and Oregon provided the international union with about 7,500 dues-paying members and with a solid achievement in its efforts throughout the 1930s to raise the banner of moderation, caution, and stability. Zellerbach wrote, "Not a single day's interruption of work has been experienced by any paper mill on the Pacific Coast," a happy circumstance due to "industry-wide collective bargaining on a regional basis and the unique methods of contract negotiations."[23]

At the same time, the international union found its locals in Washington and Oregon anything but quiescent. Throughout the decade, local unionists criticized their international union and the AFL. They sided overwhelmingly with the industrial union faction in the AFL, and after the split between the Federation and the Committee for Industrial Organization, many locals passed resolutions urging the Pulp, Sulphite, and Paper Mill Workers to join the rebel group. The very fact of uninterrupted industrial peace discredited the international union in the eyes of some Northwest activists, who reasoned that if a certain level of wages could be obtained in what sometimes appeared to be collaborative bargaining, even greater gains could be made through a more aggressive stance. Moreover, the general turbulence of labor in the region stirred the hearts of pulp and paper workers, many of whom found their organization's policies of restraint and accommodation less than heroic.

The Pacific Northwest unionists also felt that the international union took them for granted. They believed that they paid far more in per capita tax than they received in service. By 1935, John Sherman, who had earned his spurs as a local officer in Port Angeles, Washington, and had served as an international organizer, had been elected a vice-president, in effect representing the West Coast. The international union had no system of regional elections, however, and local dissidents claimed that Sherman was the choice of the international union's establishment, not of the Pacific Northwest unions. Hence unionists in Washington, Oregon, and California constantly pressed for the creation of regional vice-presidencies.

23. Kerr and Randall, *Collective Bargaining in the Pacific Coast*, 6 (quoted) and 29.

In addition to Sherman, who of course functioned as a field representative, the international union normally employed one or two additional full-time representatives for the West Coast area, which by 1938 also included two large locals in British Columbia. Periodically several special organizers were also engaged. Burke annually underwent the ordeal of a flight to the coast, usually spending several weeks in the annual negotiation sessions and visiting locals. Since some of Sherman's time was spent organizing box and converting workers in California, however, and since the other international organizers on the coast also devoted much of their time to this intricate, frustrating, and expensive task, the claims of the spokesmen for the Pacific Northwest Conference were essentially valid: the coast unions paid much more into the international union than they received in return. Although Burke constantly warned against such parsimonious calculation, Pacific Northwest locals led the fight in several biennial conventions for dues reductions. Their defeat further reinforced their judgment that they constituted an exploited minority with unique regional interests that the international union took for granted.[24]

Naturally, the West Coast locals became centers of dissidence within the international union. At intervals large locals, such as those at Longview, Hoquiam, and Everett, refused to participate in the Uniform Labor Agreement negotiations and then protested when employers, citing the terms of the agreement, refused to bargain with separate locals. Many workers in the pulp and paper mills found CIO unions such as the International Woodworkers of America attractive. In the late 1930s and early 1940s, throughout locals of the Pulp, Sulphite, and Paper Mill Workers, cells of CIO supporters sought to change affiliation. Moreover, some members, after the initial organizing enthusiasm had passed, and after industrial organization had shown itself to have some disadvantages for skilled craftsmen, sought to defect from the paper unions

24. The financial aspects of the West Coast locals' grievances are extrapolated from Auditor's Report, 1941, Reel 1. For a typical case of unhappiness on the West Coast, see, e.g., J.K. Van Buskirk, Secretary, Local 153, Longview, Wash., to Burke, April 27, 1936, Reel 3. The yearly files for locals such as 169 (Hoquiam, Wash.), 153 (Longview, Wash.), 155 (Port Angeles, Wash.), and others, and for organizer Frank C. Barnes, Jr., and Vice-President John Sherman, reflect the concerns of the Pacific locals.

and to form or expand craft organizations in the mills.[25] These were manageable problems, however, none of which seriously threatened the Brotherhood's position on the coast or impeded the smooth functioning of the Uniform Labor Agreement, which soon became famous around the country as a model of labor-management accord.[26]

If local unions were sometimes argumentative and defiant, Burke could always count on West Coast company executives for cooperation and understanding. Wolf and Burke occasionally disagreed about details concerning the Uniform Labor Agreement, of course. And once, when the travel-weary president-secretary hinted that he might forgo the annual ordeal of a trip to the coast, Wolf dressed him down, questioning his sense of responsibility. "I think you owe it to me to come out here," the corporate executive chided. But such disagreements were rare. Wolf published articles and gave speeches describing the West Coast bargaining arrangement in terms of the three branches of government. "The employees of the Pacific Coast Pulp and Paper Industry"—including plant managers and supervisors—"look upon themselves as 'citizens' of the industry," he asserted. Management, of course, comprised the "executive" branch, while the grievance procedure functioned as the "judiciary." The West Coast mills, he declared, "have order and discipline . . . , and a better *esprit de corps* than existed before the mills were organized by the two A.F. of L. unions."[27]

Despite his occasionally sour view of the indignities of collective bargaining and his distaste for the annual flight westward, Burke also regarded the West Coast arrangement as a breakthrough. In 1941 he noted, "Industrial peace has been maintained in these mills during a period of great labor unrest up and down the Pacific Coast." Burke had, he observed, "up until ten years ago, spent the greater portion of my life . . . fighting [in] strikes." The modicum of industrial peace and justice so laboriously

25. Sherman to Burke, Aug. 5, 1940, Reel 1; on the industrial union matter, see Burke to Barnes, Jan. 14, 1936, and to Sherman, May 18, 1936, both on Reel 1.

26. Kerr and Randall, *Collective Bargaining in the Pacific Coast*; Guthrie, *Economics of Pulp and Paper*, 124–32; Graham, *Paper Rebellion*, 34–41.

27. Robert B. Wolf to Burke, Feb. 6, 1936, Reel 4; Wolf to Burke, April 4, 1938, Reel 7 (quotation); typescript of Wolf, "Practical Management-Labor Cooperation," speech delivered c. March 1941, Reel 10.

achieved on the coast, while far from the perfection that Wolf believed it, indeed attested to the virtues of "fairness and common sense on the part of both the employers and labor."[28]

In the South, the other major new area of pulp and paper production, the union also relied on cautious, accommodationist tactics. Although its early efforts under the leaky umbrella of the NRA proved abortive, in the late 1930s the union began to make important gains in Louisiana, Alabama, Florida, and Georgia. By 1941, the Brotherhood together with its constant organizing partner, the Paper Makers, was well on its way to substantial organization of the southern paper industry. As in the Pacific Northwest, the unions' reputation for moderation stood them in good stead. Employers were growing apprehensive about the aggressive CIO forays into the South's textile, woodworking, and transportation industries. Most of the major employers establishing southern mills had dealt with the paper unions in the North and had come to see them, not as the incarnation of the fighting spirit of the twenties, but rather as a happy alternative to the legions of John L. Lewis.

Moreover, in a number of the new mills, employers such as Union Bag and Paper, Champion Paper and Fibre, and especially Southern Kraft brought skilled workers from the declining northern mills to man the beaters, cookers, and fourdriniers and to head the powerhouse and repair crews. These workers were often experienced union men from whose ranks came activists around whom the international unions could build.

In the 1930s the pulp and paper industry in the South grew rapidly and dramatically. More advanced chemical and cooking processes made the resinous southern pine commercially usable and encouraged investment. Eager to acquire industry, southern banks and business leaders granted tax concessions, floated loans, and otherwise courted northern paper manufacturers. Such concerns as Union Bag and Southern Kraft, a subsidiary of International Paper, needed little encouragement to look south. Ready timber supplies, abundant water sources, and the favorable physical and psychological climate made the South seem ideal to these large companies. International Paper, for example, had suffered greatly during the strike of the 1920s and had been hit hard by the depres-

---

28. Burke to Green, Jan. 30, 1941, Reel 8 (first quotation); Burke to Wolf, March 20, 1941, Reel 10 (second quotation).

sion. With its American branch all but shut out of newsprint production, and with a number of old, antiquated mills and a balky
market for the products of its book and bond division, the firm
eagerly sought to take advantage of the revolution in shipping
and packaging that required the coarse paper most easily produced
from southern pine. Beginning in the 1920s with the takeover of
small, experimental mills in the South, International Paper's subsidiary, Southern Kraft, moved quickly into the region in the 1930s,
establishing mills at Mobile; Panama City, Florida; Georgetown,
South Carolina; Camden, Arkansas; and elsewhere. A large
Champion Paper and Fibre Mill at Pasadena, Texas, several mills
of the Southern Advance Bag Company, and a giant kraft and
bag mill of Union Bag and Paper at Savannah were also erected
in the middle and late 1930s. Demand for kraft for bagging, corrugated boxes, and shipping containers leaped upward virtually
every year, so that the South became within a decade a major
pulp and paper producer, employing by 1940 approximately 20,000
workers in its new and often technologically advanced mills.[29]

Early efforts by the Pulp, Sulphite, and Paper Mill Workers to
organize southern mills largely failed. NRA-inspired locals arose in
Louisiana, Alabama, North Carolina, and Virginia. But because
the international union's efforts were focused elsewhere, southern unionists had to rely on sporadic assistance from general representatives of the AFL when intransigent pulp and paper companies in the region relentlessly discharged activists, intimidated
union members, and rejected any dealings with the unions. Nor
were the NRA labor boards of much help; they offered even less
protection to organizing workers in the South than elsewhere.[30]

The early activism in the South, however, was real. Reports from

29. Smith, *History of Papermaking*, 391–442; Bettendorf, *Paperboard and Paperboard Containers*, 31; "International Paper and Power." Northrup, *The Negro in
the Paper Industry*, gives higher figures but includes workers employed in "allied
products" (p. 34).

30. Auditor's Reports, 1934–35, Reel 1; files for Locals 36, 152, and 157, 1933–35.
Documents relating to Celotex Co., Marrero, La., and to Brotherhood Local
149, Oct. 1933–July 1934, NLB-NLRB Records, NLB, New Orleans Region, Dockets
1, 1A, and 1B; documents relating to New Orleans Corrugated Box Co., Bogalusa, La., April-May 1934, NLB-NLRB Records, NLB, New Orleans Region, Case
74; documents relating to George & Sherrard Paper Co. and International Brotherhood of Paper Makers, May 17–Nov. 6, 1934, NLB-NLRB Records, NLB, New Orleans Region, Case 61; files for Case 114 (Southern Advance Bag and Paper Co.,

beleaguered southern locals indicated a residue of union enthusiasm even as disastrous strikes and belligerent companies decimated the fledgling locals. "We are sticking as long as we can . . . ,"
reported a Louisiana activist in September 1933, "but we need some-
one with authority to lead us." By early 1936, all that remained
of the hopeful beginnings were a struggling local in Tuscaloosa,
Alabama, and an unusually energetic local with a potential of
2,000 members in Covington, Virginia. In July of that year, the
Paper Makers' organizer who had overseen much of the Pulp, Sulphite, and Paper Mill Workers' initial venture into the South had
to leave his union work. "My only regret," he informed Burke on
leaving, "is that I was unable to hold the Local Unions that we
were able to establish under the NRA."[31]

The southern situation changed rapidly, however. By 1937, several factors had come together to reshape the union's priorities
and to make employers receptive to the paper unions' initiatives.
Successful organization in other parts of the country now enabled
the union to dispatch organizers to the South, while competition
from the new mills in that section could jeopardize union wage
scales in organized mills. The CIO, fresh from its victories in rubber, auto, steel, and other mass production industries, began to
stir in the South, in turn spurring AFL national and regional functionaries into action. Not only did the specter of the CIO prod
the Brotherhood directly, it also encouraged Southern Kraft, which
quickly emerged as the leader of southern production, to soften
its attitude toward the AFL paper unions. "It is remarkable the way
the kraft paper industry is developing," Burke observed in March
of 1937. "We certainly must give attention to that southern field."[32]

The key to organizing the south plainly lay with International
Paper inasmuch as its subsidiary, Southern Kraft, dominated the
paper industry in the region. Although it had fought the unions
bitterly in the 1920s, by the mid-1930s the parent company was
ready for accommodation. In the wake of the labor turbulence
that the NRA had unleashed, the two AFL unions seemed entirely

---

La.) and Case 169 (George & Sherrard). Also File No. 176-311, Re Gulf States
Paper Co., Tuscaloosa, Ala., Aug. 1934, and File 182-209, George & Sherrard,
USCS Records.

31. Sam Moncrief to Burke, Aug. 17 and Sept. 7, 1933, Reel 2; F.W. Wenth to
Burke, July 31, 1933, Reel 3; J.C. Furr to Burke, July 10, 1936, Reel 4.

32. Burke to Schneider, March 3, 1937, Reel 7.

responsible and eminently reasonable. Burke had had cordial conversations with International Paper officials during the NRA code hearings. As the firm moved energetically into the South, and as its kraft operations became the most important profit-making arm of the giant corporation, company officials were eager to avoid labor problems.

Throughout 1937, the paper unions and International Paper officials developed an approach to organizing the company's workers. In April the parties met in Montreal and signed union shop agreements covering the operations of the Canadian International Paper Company, whose mills produced vast quantities of newsprint. The normally phlegmatic Burke was ecstatic about this development. "These negotiations were of such importance," he observed in the notations on his monthly expense report, "because from May 1, 1921 to the signing of this agreement . . . all of the mills of the International Paper Company had been open shop." With the Canadian agreement, Burke expected that International Paper's remaining plants would soon fall into line. "All that attended the [Canadian] conference were elated," he wrote.[33]

In the fall, International Paper's book and bond division, with plants in New York State, signed union-shop agreements. At the same time, Burke and Burns met with company president R.J. Cullen and Southern Kraft vice-president Maj. J.H. Friend to lay plans for organizing the southern mills. The Paper Makers proposed a "non-beneficial" plan of organizing, which entailed signing workers up but delaying the collection of initiation fees until after the union had negotiated a contract. Burke, however, rejected this approach, insisting that his Brotherhood would continue to collect its normal two-dollar initiation fee[34] and would keep building local unions around a solid core of dedicated unionists. The Paper Makers' mass organizing approach, he felt, might be necessary if the unions encountered hostile employers, "but in our paper mills where practically all of the companies will meet and deal with us, I doubt very much that this is a good plan."[35]

In August 1937, Burke appointed former Brotherhood president

33. Burke, Monthly Report, April, 1937, Reel 1.
34. One dollar went to the local union treasury; the other went to the international union and paid the new member's first month's dues of 60 cents.
35. Burke to Schneider, March 3, 1937, Reel 7.

John Malin, a veteran organizer, to lead the campaign in the Southern Kraft mills at Mobile; Panama City, Florida; Moss Point, Mississippi; and Georgetown, South Carolina. The company kept in close touch with these efforts to organize, conferring regularly with the union executives about the southern situation even while finishing up successful negotiations on the book and bond agreements. Thus, for example, in September, Burns reported to his executive board that "Friend . . . wants to confer with President Burke . . . and the undersigned for the purpose of going over the difficulties we are having in building up our local unions in the South."[36]

Malin did encounter problems. He focused his efforts at Mobile, where Southern Kraft operated a large kraft mill and a nearby bag mill, and at Panama City, Florida. Meanwhile, organizer Paul Phillips of the Paper Makers concentrated on Southern Kraft's facility at Moss Point, Mississippi. Malin, however, found that despite Phillips's previous efforts in Mobile, only a few local unionists had remained active. Indeed, it was only because the required seven workers, most of them really mechanics and machine tenders, had paid dues that the international union could properly keep Local 337's charter alive. Meanwhile, the charter of Local 229, which covered the bag mill, had lapsed entirely.

In Panama City, where he had to start a new local, Malin found that the craft unions posed problems. Impatient with the pulp workers' lethargy, the mechanics and electricians affiliated with area craft locals and sought to deal directly with Southern Kraft, thus precluding coordinated bargaining involving the various unions in Southern Kraft's plants. It had been Burke's hope that the Southern Kraft mills could quickly be organized, for once local unions were established, "I am confident that . . . we can get an agreement" covering the entire Southern Kraft operation in the South. Still, in view of the crafts' claims in Panama City, the need to start anew in Mobile, and the beginnings of CIO initiatives in these cities later in the fall, the president-secretary advised Malin to sign up the plants individually on any possible basis, for he

36. Burke to Whom It May Concern (re John Malin), Aug. 14, 1937, Burke, Monthly Report, Sept. 1937, both on Reel 1; Matthew Burns to Paper Makers' Executive Board, Sept. 15, 1937, Reel 6.

felt certain that the company would cooperate eagerly in developing subsequent companywide bargaining arrangements.[37]

Indeed, Southern Kraft plant managers provided the least of Malin's problems. They encouraged workers to join the Brotherhood locals and conferred with Malin to devise stratagems against the CIO. Malin reported to Burke in October that the manager of the Panama City plant had declared that "the Co. was pleased with the way things were being handled by you and Pres. Burns and . . . I told him how anxious our two orgs. were to continue the present relations and good faith." A month later, Malin observed that CIO activities in the southern cities could only help the union, for the local Southern Kraft manager "does not like c.i.o. so . . . I will . . . try to get him to boost [our union] indirectly."[38]

Throughout the prewar years, relations between the paper unions and Southern Kraft remained collaborative. As CIO activity mounted, the company cooperated closely with the two unions in NLRB proceedings. Moreover, the company actively encouraged workers to join the union. CIO representatives accused Southern Kraft and the AFL organizations of coercing employees, particularly low-wage black workers, into the AFL paper unions as a means of sealing off the mills to CIO initiatives. After successful agreements with Southern Kraft early in 1938, the two organizations pushed into the corporation's other mills at Camden, Arkansas; Bastrop, Louisiana; and Georgetown, South Carolina. By mid-1939, the unions had achieved union security arrangements with the company. Southern Kraft prided itself on its willingness to enforce these provisions as a demonstration of its cordiality toward the unions and as a means of helping the AFL organizations isolate and eliminate CIO dissidents.

As usual, Burke felt troubled by the union-shop arrangements, especially after he learned that other companies were criticizing Southern Kraft for its policy on union security. "It is beginning to dawn on some of these employers," Burke warned an organizer in Arkansas, "that they are making the unions for us by signing

37. Malin's reports from Mobile and Panama City are in Malin file, 1937, Reel 3; Burke to Malin, Nov. 1, 1937 (quoted) and Nov. 13, 1937, Reel 3.

38. Malin to Burke, Oct. 28 (first quotation), Nov. 25, 1937 (second quotation), Reel 3.

these union shop agreements." He felt that critics might be correct when they asserted "that the workers would not organize voluntarily and stay organized were it not for the pressure that the companies put upon them."[39]

The paper unions paid a price for the cozy relationship with Southern Kraft. Certainly, relying as they did on the enforcement of union-shop provisions and seeking to contrast themselves with the CIO, they found it difficult to present themselves as champions of the toiling masses. The depression of 1937–38 revealed the limitations of organization through collaboration. As the mills slowed and laid off workers, Southern Kraft called for union cooperation in implementing a wage cut of 5 percent. Throughout 1938, the international organizations sought to persuade angry local unionists of the force of the company's argument and to head off any effort to accompany the agreement for wage reduction with demands for company concessions with respect to grievances, work rules, and other matters.

At a meeting in Panama City on July 8 and 9, 1938, Major Friend attempted to wring voluntary agreement to the wage reduction from delegates representing six Southern Kraft locals. The delegates responded angrily, chiding the powerful corporation for seeking to extract profits from the modest pay packets of ordinary workers. While management and white-collar employees seemed to prosper, production workers bore the brunt of the depression. One delegate said to Friend, "Our debts piled up on us; we lost our radios, our electric refrigerators, our automobiles and even our homes." The delegates submitted the company proposal to the local unions for a vote; few were surprised in August when each local rejected the wage cut.[40]

On August 15, the company and the internationals made a sec-

39. Burke to A.B. Hoff, June 10, 1939, Reel 2. See also Paul Phillips to Matthew Burns, copy, Dec. 16, 1938, Reel 7. Phillips, the Paper Makers' chief representative in the South, wrote that Southern Kraft "had definitely encouraged organization of the men in the Southern Kraft mills," and Friend "had gone far beyond the contract to cooperate with us in building our organization." Phillips also hinted that Friend had helped to channel Southern Kraft funds to the AFL in the South.

40. Southern Kraft to employees, June 24, 1938, in Minutes, Panama City meeting, July 8–9, 1938, and Joint [union] Committee to J.H. Friend, July 9, 1938, Reel 7.

ond effort to gain endorsement of the wage reduction. Repeating his arguments of June, Friend warned of layoffs and permanently diminished markets unless the unionists helped Southern Kraft in this time of trouble. The local delegates again balked. "Since we . . . stalled this cut at Panama City [in July] we have taken in many members . . . ," one observed, wondering how acquiescence could contribute to further recruitment. At the very least, argued others, the unions should squeeze significant concessions on grievances, work assignments, and company policies in return for the cuts.[41]

Malin, representing the international union, would not accept the workers' argument. The place to fight grievances was not at a special conference but rather in the offices of the several plants. Fearful that some of the delegates' harsh language ("The Major does not keep his promises," asserted one delegate) would alienate company officials, Malin and Frank Berry, his counterpart in the Paper Makers, urged caution and restraint. Again and again they compared the current situation with that of 1921. The international unions had unwisely struck in 1921 over wage reductions, Barry recalled, but "we do not want to make the same mistake now." Southern Kraft had been receptive to the unions precisely because they were sober, responsible, and realistic in their willingness to recognize the conditions facing employers. The current contracts had been granted because of the good sense of the AFL unions. A small reduction on a temporary basis, carefully hedged with restrictions and qualifications, was not too much to ask of workers, who were now protected by union shop contracts and could anticipate years of prosperity and rising wages once the current emergency had run its course. Malin phrased the problem in the context of the union's history. "During the time from 1921 to 1932 not an officer of our International unions would have been permitted to enter the office of the International Paper Company discussing . . . conditions. Since that time," he told the delegates proudly, "we have practically organized entirely [the] International Paper Company." Eventually, after considerable pressure from the international unions and some rewording of the company's proposal to stress its temporariness, the local unions of both internationals accepted the reduction.[42]

41. Minutes, Mobile Conference, Aug. 15, 1938, Reel 7.
42. The remarks of Malin, Barry, and the local unionists are in Minutes, Pan-

Not all employers were as friendly as Southern Kraft. Organization of the Union Bag and Paper's Savannah mill proved more difficult, as the large and experienced company refused to follow Southern Kraft's lead in wages and union security. The Brotherhood invested almost a year of Malin's time in 1939 in an unsuccessful effort to organize a large plant of the Champion Fibre and Paper Company at Pasadena, Texas. Still, by 1941 the Southern Kraft mills were organized, and southern membership, almost nonexistent in 1937, stood near 5,000 workers, close to 10 percent of the international union's total.

A number of problems continued to plague the international union in the South. The CIO lurked on the fringes, waiting to take over AFL plants although not yet actively committed to challenging the older organizations directly. The Brotherhood's problems with the craft unions were constant. Burke doggedly attempted to deflect the imperious Machinists and Electrical Workers, caught between his sure knowledge that in any showdown the AFL Executive Council would support these powerful unions and his equally keen understanding that the paper companies resented the splintering of their workforces into craft segments.[43]

Restive local unionists, however, represented the most perplexing problem. From Burke's vantage point, southern paper workers had much to be thankful for. By 1941, they had achieved the highest rates for common labor in the region. They were represented by widely respected, thoroughly knowledgeable organizations. Many had the benefit of union shop contracts. They had made these gains in a remarkably short period without undergoing strikes, lockouts, or labor conflict of any kind. They had only to look around them to see the contrast. The CIO brought with it violence, antagonism, and confrontation. The Brotherhood, on the other hand, guaranteed steady employment, rising standards, and accord between workers and management. Yet the southern paper workers were no more grateful or satisfied than were their counterparts elsewhere.

---

ama City Conference, July 9, 1938, Reel 7; the final agreement is contained in Paul Phillips to Friend, Aug. 27, 1938, Reel 7.

43. Membership figures are extracted from Auditor's Report, 1941, Reel 1. On problems with the craft unions, e.g., Burke to H.W. Brown, Vice-President, International Association of Machinists, Nov. 23, 1938, and related material, Reel 7.

Particularly perplexing were the black workers.[44] Despite the union's earnest efforts, and despite real gains in wages, the Negroes, too, often proved apathetic. Worse, in some places they seemed uniquely susceptible to the CIO appeal. As Burke told A. Philip Randolph, although the organization's dues and initiation fees were modest and although "colored workers" had benefited from Brotherhood-negotiated contracts, "for some. reasons I could not understand, the colored workers hold aloof from joining our Union." Of the 5,000 or more southern members in 1941, about 750 were black, many of them enrolled in segregated Brotherhood locals through union shop provisions and left with little representation in union affairs and little protection.[45]

If black workers were a disappointment to Burke, whites too often proved impatient, unrealistic, and fractious. "They will . . . have to learn that business conditions govern what we are able to get in the way of wages," Burke declared. Local unionists, however, often balked when they were told to feel grateful. During negotiations in Camden, Arkansas, in 1939 an international organizer sought to impress local unionists with the Pulp, Sulphite, and Paper Mill Workers' long record of accomplishments. He ran into an impatient worker with a vivid, if slightly askew, historical sense. "I am getting damn sick and tired of hearing what the West Coast . . . has done," this dissident proclaimed. "We don't care what happened in Grover Clevelands [sic] time."[46]

But despite these sporadic problems, the effort to organize the South was a clear success. In 1936, when he first contemplated moving forcefully into the South, Burke said, "It is not going to be a very pleasant job."[47] Yet he proved wrong. Although the two paper unions did not achieve as regionally integrated a bargaining structure as that which emerged in the Pacific Northwest, and although CIO challenges and racial tensions remained, the breakthrough with International Paper and Southern Kraft and the

44. Northrup, *The Negro in the Paper Industry*, 31–39.
45. Estimates of membership are based on Auditor's Reports, 1941, Reel 1; Burke to Charles Miller, Local 388-A, Savannah, Jan. 31, 1941, Reel 6. See Northrup, *The Negro in the Paper Industry*, 32–34, for a different approach to employment figures.
46. Burke to Hoff, June 3; Hoff to Burke, June 8; Burke to Hoff, June 10; Hoff to Burke, June 15, and subsequent correspondence, 1939, Reel 2.
47. Burke to A.M. Buck, Sept. 3, 1936, Reel 2.

rapid organization of the new mills went far beyond the hopes of 1936. With 5,000 southern workers organized, with energetically enforced union shop contracts prevalent in the region, and with a record of substantial recent wage increases for the international union to claim credit for, by 1941 the task of organizing the South had proved an agreeable one indeed.

Building the union in the South and the Pacific Northwest was a neat and relatively painless process by comparison with movement into the converted paper segment of the industry. In the period before World War II, the union moved forcefully into this diverse, sprawling sector after an uncertain beginning in the NRA era. With the rise of the CIO threat in 1937, Burke and his colleagues rapidly reassessed their diffident attitude toward the relatively unskilled, ethnically diverse, and scattered converted paper workers, many of whom were women, and launched a series of organizing campaigns that centered on major urban areas. Moving from a foothold in New York City, the Brotherhood extended operations into Philadelphia, Pittsburgh, the greater Boston area, and other centers of paper converting. Although the union suffered some serious setbacks — its efforts in Chicago fell flat, and its campaigns gave way before those mounted by the International Printing Pressmen in Southern California — by 1941 it had organized at least 15,000 converted paper workers who accounted for at least 25 percent of the international union's total membership.

Rights to organize box, bag, novelty, and sundry paper workers had been implicit in the international union's charter since 1909.[48] The jurisdiction was not unambiguous, however: it overlapped with claims that unions in the printing trades might legitimately make. Nor did the beleaguered Brotherhood move aggressively into the field. In 1917 the international union chartered a New York City local for workers in the paper bag and paper plate trade, but throughout the 1920s and early 1930s the grim struggle to survive in primary pulp and paper consumed all of the organization's energies. Indeed, in the mid-1920s, Burke virtually waived claims to workers in the box industry, concluding that "this field is a little outside of the proper sphere of activity for the Pulp, Sulphite and Paper Mill Workers." He advised New York City box workers

---

48. Brotherhood, *Constitution and By-Laws.* Jurisdiction was formally expanded by action of the 1922, 1926, and 1935 conventions.

to seek a federal charter, for "we have not the time or the means to devote [attention] to the paper box industry now."[49]

By the 1930s, however, the union could no longer afford to take such a detached attitude. All observers agreed that converted paper was rapidly outstripping primary pulp and paper in terms of investment, value of product, and employment. Moreover, while the industry continued to be decentralized and largely geared to the peculiarities of local products, large paper manufacturers had begun to build integrated facilities providing a wide variety of packaging goods to dairies, bakeries, and wholesalers on the outskirts of major urban centers. Primary pulp and paper remained geographically tied to timber and water resources, but the converting industry increasingly gravitated toward its markets, so that by 1940 there were important concentrations of workers in the key states of New York, Massachusetts, Pennsylvania, Illinois, and Ohio. New York, for example, had about 550 plants employing almost 28,000 workers; both Pennsylvania and Massachusetts held more than 200 plants and more than 12,000 employees. New Jersey, California, Michigan, and Indiana were also among the larger converted paper producers. In 1939, the Census of Manufacturers counted almost 3,500 converting plants in the country, employing 126,570 workers. Department of Labor figures for both were higher.[50]

Prior to the 1930s, few people associated with the Pulp, Sulphite, and Paper Mill Workers had had contact with the converting sector. Veteran unionists were constantly perplexed and even awed by the bewildering variety of products and the wide range of problems that employers and workers in the industry encountered. Although they prided themselves on their knowledge of their trade, Brotherhood representatives could not always keep pace with the dynamic packaging industry. "There is plenty to learn about the paper box industry," Burke told an organizer in 1937, adding, "It

49. Burke to Frank Morrison, April 28, 1925, and Brotherhood, Executive Board, to Fred Caiola, March 13, 1926, AFL-CIO Records (AFL-CIO Building, Washington), Reel: National 12: Brotherhood.

50. CIO Staff Memorandum, Feb. 3, 1944, summarizing census and other data, AFL-CIO Records (AFL-CIO Building, Washington), Reel: CIO National 2: Paper Workers of America. Most other estimates give at least 200,000 as the correct figure for employment in paper converting; Burke, Report, *Report of Proceedings . . . 19th Convention*, 1941, p. 83.

is no cinch to represent the workers in these paper converting plants, because so many different products are made and the mechanical processes are so different."[51]

Adding to the organizers' difficulties was the wide variation in the size and character of manufacturing facilities. Modern plants erected by the Container Corporation of America, National Container, and other emerging converting giants were large, mechanized, sophisticated enterprises, with closely integrated paper supplies, transportation facilities, and urban consumer markets. Concurrently, the set-up box industry, geared to local special products in large cities, remained important, particularly in employment, as well as relatively labor intensive. As a result, when unionists spoke of paper converting, they had to envision an industry that embraced large, modern facilities employing more than 1,000 workers and small, crowded shops employing fifteen or twenty women in cramped lofts and basements. One detailed CIO report placed the average number of workers in a converted paper plant at about sixty-five but noted that many firms were smaller or larger. Thus organizers might have to erect complex, multiemployer, citywide bargaining arrangements in the set-up industry, involving hundreds of small shops while at the same time dealing with the large converting plants integrated into the increasingly complex urban transport, packaging, and food-handling networks.[52]

The Brotherhood first exercised its jurisdiction in converted paper in a substantial way after the advent of Section 7(a). While Burke and his organizers concentrated on primary pulp and paper and on the creation of the Pacific Northwest bargaining structure, paper box, bag, and novelty workers sought out the union, referred to Fort Edward by AFL functionaries and craft unionists who assisted in the early surge of organization. A wave of organization swept through the box and bag shops of Ohio. Several converted paper locals sprang up in Milwaukee, while box, bag, and novelty workers in southern New England and New York City formed locals in these exuberant months. Much of this organization proved ephemeral, however, because of the uncertain guar-

---

51. Burke to Frank C. Barnes, Jr., July 23, 1937, Reel 1.
52. CIO Staff Memorandum, Jan. 29, 1944; National Paper Box Manufacturers' Association to Wayne Lyman Morse, Chairman, Industry Committee 14, Oct. 10, 1940, AFL Papers, State Historical Society of Wisconsin, Series 4, Box 97.

antees provided by the NRA, the inexperience and discouragement
of converting plant workers, and the inattention and impatience
of the international union. By the end of 1935, only a few converted paper locals remained of the dozens that had been created
in 1933 and 1934. The union still held on in Toledo and Milwaukee, retained scattered locals in New England, and was kept active in New York City through the tenacity of Local 107, originally chartered in 1917. But by the end of 1935, converted paper
workers accounted for only about 6 percent of the organization's
15,000 members.[53]

The attitude of Burke and the other veterans toward converted
paper workers was always ambivalent. On the one hand, they realized that the future of the whole paper industry lay not in the
pulp and newsprint mills, or even in the detached paper-making
plants, but in the converting sector. And Burke believed that "it
requires a union like the Pulp, Sulphite and Paper Mill Workers
to really organize the thousands of low-paid workers in this industry." At the same time, he could not suppress a certain disdain for these less-skilled, urban, and often ethnic workers. Even
as converted paper came to take on more and more importance
within the international union, Burke idealized the union's old
constituency and contrasted the behavior of converted paper workers unfavorably with that of the pulp and paper mill stalwarts.
He endlessly complained about the inexperience, fecklessness, and
ignorance of box, bag, and novelty workers. "I doubt if we have
a single local union in a converting plant that is an asset to the
International Union," he remarked in July 1936.[54]

Part of the problem with the converting plants was that so many
of the workers in the urban areas had ethnic backgrounds and
temperaments differing from those of the veteran but often provincial international organizers who dealt with them. Moreover,
the union's experiences with urban locals in the past led Burke
and his associates to suspect that deep involvement in the inter-
union machinations of, for example, the New York City labor
movement would associate the international union with chaos,
corruption, and communism. Representatives in the city believed

53. Membership figures are gleaned from Auditor's Reports, 1936, Reel 1.
54. Burke to Fred C. Hoodwin, Dec. 22, 1939, Reel 7 (first quotation); Burke
to Jacob Stephan, July 3, 1936, Reel 2 (second quotation).

as an article of faith that the truck drivers' local of the International Brotherhood of Teamsters regularly extorted money from paper companies for transporting and handling materials coming into the city's set-up and converting plants. Organizers sent to northern New Jersey reported that various unsavory characters had followed them, letting them know that "established" labor groups in the area did not welcome an initiative on the part of the Brotherhood. [55]

Urban converted paper locals were also a source of the little radical influence that existed within the Brotherhood. Hyman Gordon, president of New York City Local 107, was a vocal leftist, constantly pressing the international union to adopt confrontationist tactics with regard both to employers and to the AFL establishment. As Burke himself acknowledged, Gordon and his followers were strong union activists, supporting picket lines throughout the city and eager to extend the organization of paper workers in the metropolitan area. At the same time, he regarded Gordon as a follower of the Communist party line and as an incipient Stalinist whose trade unionism, in the final analysis, took a back seat to his ideological commitments. [56]

Even those urban unionists who were neither openly radical nor tainted with corruption were vaguely suspect. The Jewish and Italian unionists prominent in New York paper organization appeared aggressive, sharp-tongued, and disconcerting to organizers whose lives had been spent in the small villages and mill towns of New England and New York State. Edward Mangan, no doubt passing along information gleaned from old-timers in the union, remarked that the international union had once organized some New York City workers, "but the membership was composed mostly of Jews and was hard to please and we had given them up." At a 1936 strike of a bag workers' local in the Bronx, Burke apprehensively envisioned the chaos that would arise when the various paper industry unionists descended upon the scene to help the bewildered pickets. Gordon would lead a cadre of radicals from Local 107 and would seek to instruct the strikers in the dialectics

55. Raymond Leon to Burke, Jan. 11, 14, c. Feb. 2, 1936, Reel 1.
56. See, e.g., Burke to Dennis J. O'Connell, Jan. 15, 1936, Reel 3. Copies of Local 107's militant newspaper, *Paperworker Champion*, are in Brotherhood Papers, 1938, Reel 5.

of the class struggle. Meanwhile, the Paper Box Makers' Union, a federal union formerly aligned with the Pulp, Sulphite, and Paper Mill Workers, would send its business agent, a bustling, fast-talking little man always bursting with projects and ideas. No doubt, Harriet and Morris Wray, a brother-and-sister team, would appear as well, adding their earnest radicalism to the scene. "Jake [Stephan] and the brothers and sisters are picketing the factory," he reported to Vice-President Sullivan. "Brother Gordon is also over helping . . . and Brother Weinberg has been up and Miss Wray has joined the melee and so they are having a real good time down there."[57]

In April Stephan had another disconcerting encounter with New York City's eccentric paper unionists, for he was the Brotherhood's representative at a testimonial honoring Hyman Gordon, Local 107's radical president. Stephan, whose natural habitat was the mill towns of New York State, reported everything in detail to Burke—the outrageous charges for hat checking, the easy intermingling of races and sexes, the scandalous drinking habits of these urban, Jewish, radical unionists. "They had plenty of beer and booze and you should have seen the officials of the communist party, both men and women, drink up the booze. I tell you they drinked it by the *watterglass* [*sic*] full at a time," reported the veteran organizer, aghast.[58]

Thus, after the NRA enthusiasm had worn off, and before there were impelling reasons to move aggressively into converted paper, the Brotherhood remained diffident. Box makers and bag workers properly belonged to the international union's jurisdiction, and if the workers were patient and responsible, the Brotherhood would see to their interests. At the same time, these workers often proved more trouble than they seemed worth. "If we had all the paper box factories organized into our International Union," Burke declared in 1935, "I would, undoubtedly, end my days in a lunatic asylum." As usual, George C. Brooks was more blunt. He "would rather take a chance on most any kind of labor than paper box workers," he declared, for "the vast majority are just a bunch of grinning idiots."

57. Mangan to Burke, Jan. 23, 1936, Reel 2 (first quotation); Burke to Sullivan, March 27, 1936, Reel 1 (second quotation).
58. Stephan to Burke, April 19, 1936, Reel 2.

In August 1936, Burke sent Stephan to Essex and Passaic counties in New Jersey, at the behest of a New York City unionist who discerned vast organizing possibilities. Stephan's instructions were to assess the situation and to see if the optimistic report merited a commitment of the international's resources. But, the president-secretary warned his trusted organizer, Stephan should not hesitate to abandon the project if workers proved unresponsive or, as Burke clearly anticipated, "if you find that there is no good material and they are a bunch of crooks or Communists."[59]

With the emergence of the CIO in the first half of 1937, however, Brotherhood leaders could no longer afford the luxury of such ambivalent attitudes. While the CIO did not directly threaten a massive campaign directed at paper workers, it quickly became clear that the enthusiasm that the Flint sit-down, the success in steel, and other CIO initiatives had generated would spill over into all areas of labor. In particular, CIO activism in and around the major cities posed a direct challenge to the small, fragmentary, and haphazard urban locals of the Pulp, Sulphite, and Paper Mill Workers. While the CIO had no organization as yet for paper and box workers, the dissident federation's representatives showed little regard for the niceties of jurisdictional demarcations and soon evinced a willingness to sign workers up helter-skelter, without regard for the claims of AFL unions that had often failed to exercise their jurisdictional rights.

Thus, during the spring of 1937, Burke made a sustained commitment to organizing the paper box industries of major cities. He launched drives in Philadelphia and Chicago and took steps to make the union a significant force in the New York City paper industry. For years, a federal labor union of paper box makers in New York had dickered with the international union for affiliation. Led by business agent A.N. Weinberg, Federal Labor Union 18239 had been organized in 1933 and had sought with mixed success to bring together the scores of small set-up box shops in the city and to provide order and stability to a trade that was sharply competitive, declining in importance, and characterized by poor wage rates and working conditions. Burke believed the box mak-

59. Burke to Weinberg, Dec. 4, 1935, Reel 3 (first quotation); George Brooks to Burke, Oct. 24 (second quotation) and Sept. 28, 1935, Reel 1; Burke to Stephan, Aug. 5, 1936, Reel 2.

ers' union to be characterized by disputation, bickering, and fu-
tility. For several years he resisted Weinberg's efforts to lead the
federal union into the Brotherhood. "The less I think of the paper
box makers the better it is for my peace of mind," he declared in
1936. Yet, a year later he was forced to think of them, and in the
spring he negotiated an arrangement with Weinberg to have Federal Labor Union 18239 transferred into the international organization as Local 299. "I have decided to issue a charter to them," he
noted, but "I am doing so without any enthusiasm." Since the
international union had begun organizing box workers in Philadelphia and Chicago, however, "and also in view of the fact that
the c.i.o. now has a strong foothold in New York, we cannot afford to let the c.i.o. invade this industry." Thus, he concluded with
more resignation than enthusiasm, "we must take the responsibility of organizing the entire industry, come what will."[60]

At first, the international union concentrated on the East Coast
and in Chicago. In the Boston area, international auditor Brooks
organized box, tube, and container workers in Fall River and New
Bedford, Massachusetts, in Providence and Pawtucket, Rhode
Island, and in the greater Boston area. New England AFL officials
spurred this initiative, mobilizing the boycotting and picketing
power of the craft unions against any possible CIO initiatives.[61]

Philadelphia was a particularly important focus of organizing
activity. In April an AFL unionist from the city sent Burke a lengthy
report on the paper trades in the city, noting that there were more
than 5,000 paper box workers and that "Phila. is ripe for organization at once." Burke responded quickly, dispatching Raymond
Leon to Philadelphia to begin the task of creating organizations
among the various categories of box workers.[62]

Throughout the spring and summer of 1937, Burke pressed the
campaign. He sent at least three additional organizers into the
city. One of the early benefits of the affiliation of the New York

60. Burke to Raymond Leon, June 30, 1936, Reel 1 (first quotation); Burke to
Harry Beach, June 1, 1937, Reel 2. For Burke's criticisms of the box makers' past
behavior, Burke to Hyman Gordon, July 6, 1935, Reel 3.
61. Brooks to Burke, May 2, 4, 11, 13, 19, and other Brooks correspondence
through the summer of 1937, Reel 1; Harry P. Grages to Burke, June 28; Burke
to Grages, July 12; Burke to Grages (telegram), Sept. 22, 1937, Reel 7.
62. Joseph L. Ziegler of the Hotel and Restaurant Workers to Burke, April
12, 1937, Reel 2.

City box makers was the availability of organizers associated with the federal union familiar with the intricacies of the set-up box trade. He wasted no time in sending Morris Wray, a veteran New York unionist, to assist Leon, whose previous experience had been largely confined to the traditional paper industry. "Because of the peculiarities of the structure of organization as concerns a local union in a city where there are several shops small and large," Leon reported, "I must . . . take orders from Mr. Wray because he formed the New York local." The Philadelphia situation, he went on, "differs a lot from the one mill local in the paper industry." By the fall of 1937, Philadelphia Local 286 had signed several dozen contracts with employers in the competitive set-up box trade and was well on the way toward the creation of a multiemployer bargaining structure for the industry. Indeed, employers were encouraging the union to organize outlying shops and converting plants so as to eliminate wage differentials and hence to stabilize competitive conditions. Organizer Lambert Louy reported in November, "The box makers association has even offered us financial assistance in the matter."[63]

At the same time, Burke initiated organization in Pittsburgh and Chicago. In June he sent organizer Harry Beach to the New York metropolitan area, with vastly different instructions from those diffidently given to Stephan a year earlier. "I think the proper strategy now is to start a good strong organizing drive in both New Jersey and New York," he declared. Beach quickly reported optimistically on the prospects of organizing 1,600 set-up workers in Newark and large numbers of additional converting plant workers in Jersey City, Bloomfield, and elsewhere in northern New Jersey. With stable organizations already existing in Milwaukee and Toledo, and with some plants that had been organized since the NRA days in Cleveland and northeastern Ohio, by the end of 1937 the Brotherhood had made substantial headway in large urban converted paper fields. While membership in these areas had rarely risen above 500 through 1936, by the end of 1937 the international union counted at least 5,000 converted paper work-

---

63. Leon to Burke, June 6, 1937, Reel 2, and Leon's correspondence from Philadelphia throughout 1937; Lambert Louy to Burke, Nov. 8, 1937, Reel 2 (second quotation).

ers, the large majority of them urbanites and about 40 percent of them women.[64]

Growth in this field continued throughout the period preceding World War II. The union centered its efforts on the large converting mills near metropolitan areas to supplement its organization of box workers in the cities. Converted paper membership continued to grow almost month by month. Although there were only about fifteen converted paper locals at the end of 1935, about half of them barely maintaining their charters, by the end of 1938 there were at least fifty-three, including some of the largest and most financially stable in the international union. By the end of 1941, the number of locals had grown to about seventy-five, and converted paper membership stood at close to 15,000, fully 25 percent of the organization's total.[65]

Rapid organization of the converted paper industry did not occur without problems and setbacks. Organizers in every locale had their discouraging moments, for even in the charged atmosphere of the late thirties, not all workers flocked to the union banner. North Jersey in particular proved difficult to organize; the converting industry was highly diverse in that state, and its workers represented a bewildering variety of ethnic groups. Northern New Jersey's converting plants supplied an equally bewildering variety of manufacturers and packagers of all sorts of consumer goods, industrial components, and construction commodities. Although local unions were established in Hoboken, Jersey City, Bogota, and other aging, dingy New Jersey cities, organizers always found the area both discouraging and frustrating. Organizer George Cook, who had taken part in the third of the international union's three major efforts to build and consolidate organization in the area, reported late in 1941 on the difficulties of organizing in Newark. His day was an endless series of furtive meetings with employees of scores of small shops in the city's port area, as he sought in vain to persuade them to unionize. But Cook could report only "insults from men and women, who in their own hearts,

64. Burke to Harry Beach, June 15, 1937 (quotation), and correspondence with Beach throughout the summer of 1937, Reel 2. Membership estimates are drawn from Auditor's Reports, 1936 and 1937, Reel 1.
65. Auditor's Report, 1941, Reel 1.

know on pay day they do not receive the amount of pay they should."[66]

Nor was the international union ever able to organize the large Chicago paper box industry, despite repeated efforts going back to 1936. Agnes Nestor of the Women's Trade Union League of Chicago collaborated with Burke in developing organizing strategies to appeal to the heavily female labor force. Weinberg, always ready to instruct the president-secretary, urged massive organization in the nation's second largest city. In April 1937, Burke sent organizer Walter Trautmann, who had been successful in organizing and maintaining a half dozen Milwaukee converting locals in the NRA period, into Chicago with instructions to "give these other unions a run for their money," a reference to the possibilities of CIO incursions and the machinations of AFL printing trade organizations claiming box workers.[67]

Alas, the Brotherhood did not succeed in Chicago. The Republic Steel killings of May 30, 1937, had an intimidating effect on labor organization throughout the Chicago area during the summer. In addition, Burke's representatives reported that Chicago box makers were unusually obdurate, freely resorting to intimidation, harassment, and selective favoritism. The union's drive in Chicago completely stalled. The only local left by 1939 was a bogus union concocted by a huckstering lawyer who sought personal financial gain by means of a phantom charter. The only positive result of the initiative in Chicago was the absorption of a large federal union in nearby Morris, Illinois, which eventually added more than 1,000 members to the international union.[68]

Nor did the Brotherhood make headway in Los Angeles or the Bay area, the two centers of converted paper production on the West Coast. Throughout the late 1930s and early 1940s, Burke

---

66. Cook to Burke, Dec. 15, 1941, Reel 6.

67. Burke to Walter Trautmann, June 6, 1937, Reel 3.

68. For the Chicago situation, see Burke to Trautmann, July 6, Oct. 27, 1937, Reel 3, and other communications in this file; Valeria Brodzinski to Burke, Nov. 4, 1937, Reel 2; and Leonard C. Bajork, Regional Director, Chicago Region, to Nathan Witt, NLRB Executive Secretary, Nov. 20, 1937, and related communications, 1937–39, NLB-NLRB Records, NLRB-2 Case XII-C-631, and election results, March 17, 1938, Case 483. On the phantom local: File for Local 376, Brotherhood Papers, 1939, Reel 3. On the Morris success: Trautmann to Burke, May 25, July 27 and AFL organizer Alexander Marks to Burke, May 20, 1937, Reel 3.

hoped to duplicate the successes in Philadelphia and New York in southern California, keeping organizer Frank C. Barnes, Jr., in Los Angeles for months at a time, dispatching Vice-President John Sherman periodically, employing temporary special organizers, and even setting up an office in Los Angeles. While the international union had some success, launching large converting locals in Southgate and Vernon near Los Angeles and establishing several locals in and around Oakland, organization under its aegis remained fragmentary and lacked vigor. Burke never seemed quite to grasp the importance of Southern California, still apparently regarding it as a quasi-developed, secondary locale. "You know," he mused to Barnes, who constantly pleaded with Burke to allocate additional resources to the Los Angeles area, "California is one state that I have never been in," although "I have heard a lot about it."[69]

Still, the presence of aggressive and effective organizers from the International Printing Pressmen's Union on the West Coast proved to be the decisive factor in the converting plants there. Although Sherman fired off warning after warning to Burke that the Printing Pressmen "are spending enormous sums of money and [are] using other tactics so that they may gain the entire paper converted industry on the Pacific Coast," Burke remained calm. Preoccupied with bitter struggles against the CIO and AFL craft unions in New York City, he remarked only that "the pulp and paper and the paper products industry is so scattered and is so varied that it would be utterly impossible for our International Union to prevent other Unions from making some inroads into our territory."[70]

Burke shrugged off organizing setbacks in converted paper. When Mangan reported the Printing Pressmen active in Baltimore, the president-secretary merely noted that there remained so many box workers to organize that "our best plan is to follow the lines of least resistance."[71] Moreover, the veteran labor leader never entirely freed himself from his prejudice against converting workers, repeatedly claiming that they cost the international union much more in organizers' salaries, strike benefits, and other costs

69. Burke to Barnes, July 6, 1937, Reel 1.

70. Sherman to Burke, Feb. 12 (first quotation); Burke to Sherman, July 24, 1941 (second quotation), Reel 1; interview with Cornell University professor George W. Brooks, Feb. 7, 1979.

71. Burke to Mangan, Sept. 26, 1941, Reel 3.

than they returned to the treasury. "I know of no group of workers who need organization much more than do the workers employed in the paper converted products industry," he observed in 1941, adding, however: they "are among the most difficult [to service and satisfy] in the entire trade union movement." By 1941, some of the largest unions in the international represented converted paper workers. Local 299 of New York City paid a per capita tax of about $1,000 each month, while Local 286 of Philadelphia added about $700. Yet Burke persisted in believing that if it were not for the dues collected from the older locals in the primary paper field, the Brotherhood could not afford the luxury of maintaining a presence in the converting sector. "I sometimes wonder," he confessed to his old colleague, S.A. Stephens, "what we are going to do with these converting paper locals."[72]

Even so, he boasted of the union's accomplishments in the field. He and Vice-President Sullivan repeatedly asserted the union's primacy in converted paper in communications and depositions before the Department of Labor's Wages and Hours Division, seeking a major role for their organization in that agency's deliberations. The bargaining structure in the New York and Philadelphia areas grew increasingly sophisticated, as the international union, prodded by representatives of these urban locals, developed multiplant and multiemployer units. In 1938, representatives of New York, Philadelphia, and New Jersey locals began a series of conferences to encourage further organization, to exchange information on standards, and to reaffirm the solidarity of box workers. Led by Local 299 activists, the first conference at the end of April ended with a reception sponsored by the New York City locals and with the cry "Long live the Paper Box Makers' Unions! Long live the International Brotherhood."[73]

On a less celebratory note, the converted paper workers of New York did pose problems for the international union. There was

72. Burke to John Peters, March 24, 1941, Reel 7 (first quotation); Burke to S.A. Stephens, Jan. 30, 1941, Reel 1 (second quotation). Also, Burke to Stephens, Feb. 6, 1939, Reel 1, and to Joseph Burke, May 6, 1939, Reel 3. Receipts from the various locals are revealed in Auditor's Reports for each year, Reel 1.

73. Minutes in Anne Emaniela to Burke, May 4, 1938; Minutes, Box Makers' Conference, June 19, 1938, both on Reel 5; Brotherhood *Journal* 22 (May-June 1938), 8 (quoted); Herbert W. Sullivan, Statement, Brotherhood *Journal* 24 (Nov.-Dec. 1940), 5–7; Burke to John Sherman, Aug. 19, 1940, Reel 1.

little love lost between Local 299 and Hyman Gordon's left-wing Local 107. In turn, international officials viewed both groups with a certain skepticism, preferring to rely on Joseph Tonelli, who came out of Bronx Local 234, a bag makers' organization, for key organizing tasks in the city and its environs. The various organizations quarreled over the jurisdiction demarcations in the metropolitan area and over the virtues and defects of multiplant, as opposed to single-plant, locals. Burke and his wife were personally close to Harriet Wray of Local 299, but he regarded Weinberg, its business agent, as something of an enthusiast and Miss Wray and her brother Morris as uncomfortably left-wing. While crediting Gordon with considerable services to the cause of labor and while acknowledging Local 107's willingness to organize, support picket lines, and provide union stalwarts, Burke did not care for Gordon's strongly pro-cio attitude or for his penchant for ideological proclamations. Hence Tonelli, who was always well-informed, never argumentative, and always eager to put himself under the tutelage of the international union, emerged as the most reliable, and in the view of Burke and other members of the executive board, the most effective of the urban organizers from the turbulent New York City scene. In 1940 and 1941, when cio organizations made serious efforts to move into the corrugated box industry in the city, it was Tonelli who earned the task of fighting the "Communists," as Tonelli called the cio activists who hurled insults and more tangible items at Brotherhood unionists in a score of box shops and corrugated plants.[74]

By the eve of World War II, the Pulp, Sulphite and Paper Mill Workers, which had been virtually shut out of converted paper in 1935, was enormously successful in the field. Locals 299, 286, and 381 (Greater New York Folding Box Workers) all had multiemployer agreements, as did other locals in Minnesota, Ohio, and Pennsylvania. "I am very pleased with the continued progress in Philadelphia and vicinity," Burke informed Morris Wray in March 1940, adding, "We are also making very good progress in New York and New Jersey."[75]

74. Burke to Tonelli, Oct. 23; Tonelli to Burke, Oct. 24, 1940, Reel 3; Burke to S.A. Stephens, Feb. 21, 1941, Reel 1; Tonelli file, 1941, Reel 3. See also file for Local 107, 1940, Reel 6, and for 1941, Reel 5.
75. Burke to Morris Wray, March 21, 1940, Reel 3.

Setting aside for the moment his doubts about the steadfastness and good judgment of converting plant workers, late in 1939 he greeted the members of Local 286 of Philadelphia, rapidly on its way to being the international's largest local. "A story could be written about our campaign of organization among the low paid workers in these paper converting plants," he commented, ignoring for the moment the fact that the initiative had come from the shops. "The workers in scores and scores of these . . . factories have been transformed from timid, browbeaten, exploited human beings into self-respecting American citizens," largely through the efforts of the international union.[76] There had been failures in Chicago and Los Angeles; the CIO continued to be a challenge in New York; and it was a disconcerting, if little acknowledged, fact that converting workers and the leadership cast up from their ranks might one day have greater influence than the union's first supporters in the pulp and paper mills. Still, the Brotherhood's success in the converted field was the most dramatic aspect of the organization's revitalization in the 1930s.

Throughout the late 1930s, the union enjoyed success in each of its major organizing efforts. It reorganized its old northeastern locals and greatly enhanced its organization in the lake states. It worked effectively to make the Uniform Labor Agreement on the West Coast smooth running and strike free. In the South, that graveyard of laborite hopes, it had achieved union shop contracts in some of the largest and most modern mills. And it was well on its way to overcoming the ethnic and sexual divisions that impeded the organization of converted paper.

The Brotherhood achieved these gains largely through a policy of moderation, by associating closely with employers where possible, and by projecting an image designed to contrast sharply with that of the hungry, radical, and aggressive CIO. To those who might assert that it rode to success on the wave of CIO militancy, Burke responded that, on the contrary, it was his union's long record in the industry, its courageous struggles in the 1920s, and its willingness to seek realistic solutions in the troubled thirties that provided the key to success. With its revival of unionism in the Northeast, its innovative role in the Pacific Northwest, its pio-

---

76. Burke, Statement of Welcome, quoted in Local 286 report, *Brotherhood Journal* 24 (Jan.-Feb. 1940), 9–10.

neering victories in the South, and its growth in bag, box, and container shops around the country, it could claim to have negotiated scores of union shop contracts covering all manner of workers in one of the nation's largest industries. To Burke, these achievements represented a victory for industrial unionism that stood in sharp and favorable contrast to the fractiousness, violence, and absence of comprehensive contracts that accompanied the rise of the much-publicized cio.

7

A Union among Unions

THROUGHOUT THE TEMPESTUOUS 1930S THE INTERNATIONAL Brotherhood of Pulp, Sulphite, and Paper Mill Workers constantly faced problems as it sought to define its role in the broader American labor movement. Both high principles and jurisdictional disputes constantly restricted its vision of itself and its place in the spectrum of American labor. Although they were loyal to the AFL, union leaders nonetheless frequently criticized Federation policies and practices. In addition, because theirs was an industrial union with a broad jurisdiction, the Brotherhood continually collided with AFL craft unions, both in the pulp and paper mills and in the converting shops. It was a strong supporter of the industrial union position at the AFL conventions of 1934 and 1935 and sympathized with the early activities of the Committee for Industrial Organization. By the late 1930s, however, the organization was locked in sharp conflict with its erstwhile industrial union brothers, now in the rival federation. Both the victim and the legatee of the fierce rival unionism of the 1930s, it attempted to organize the industry as a loyal affiliate while seeking to prod the Federation toward a more expansive view of its relationship to mass production workers.

The Pulp, Sulphite, and Paper Mill Workers remained true to the American Federation of Labor. Its origins lay in federal labor unions that the AFL had chartered in the early years of the century. It had gained affiliation as an international union in 1909. Failure to build a powerful organization in the pulp and paper industry prevented the union's leadership from claiming great influence within the AFL, and by clinging to industrial unionism and to mild socialism, Burke and his associates compounded their isolation from the Federation's centers of power. Still, they believed fervently that the old Federation was *the* house of labor in the United States. Other, more "radical" organizations had come and gone – the IWW, the One Big Union in Canada, DeLeon's Socialist Trades and Labor Alliance. But now, Burke observed, "they are gone and all but forgotten." Enduring, repelling the attacks of Left and Right, and upholding the American laborite tradition was "the conservative, slow, out of date A.F. of L. . . . still here, plugging away."[1]

Expediency also compelled loyalty to the AFL. As a small union, the Brotherhood often depended on the AFL infrastructure – on the state federations, on the city centrals, even on the craft unions – for organizing help, particularly in the early stages of its activities in unfamiliar locales. As an organization without its own specialized bureaucracy, the union often called on the AFL's research division, its legal staff, and its legislative and lobbying activities. Most importantly, however, it adhered to the AFL because its leaders feared destruction if they strayed. If the Pulp, Sulphite, and Paper Mill Workers cast in their lot with dissident elements, Burke believed, the voracious craft unions would everywhere devour his locals. Defection from the AFL, he remarked in 1936, "would bring down upon us such a horde of craft unions seeking to pull us apart that it is a question if we would be able to survive."[2]

However loyal, Brotherhood leaders were far from uncritical. Throughout the 1930s Burke castigated the Federation repeatedly. He spoke out against the more imperious tendencies of its officialdom, criticizing the high salaries and elaborate perquisites voted to top functionaries at the annual conventions. He spoke out against the AFL Executive Council on matters of policy, sharply criticizing it for foot dragging and lack of responsiveness in the NRA period on the question of industrial organization. He also

---

1. Burke to Shirl N. Montgomery, April 23, 1938, Reel 4.
2. Burke to John Sherman, Aug. 12, 1936, Reel 1.

dissented from its attacks on the National Labor Relations Board
in 1938 and 1939. He criticized its subservience to the craft unions,
defended the early dissidence of the CIO elements, and upheld CIO-
secured gains even after the decisive break.[3]

By far the most serious conflict between the Pulp, Sulphite, and
Paper Mill Workers and the Federation concerned industrial unionism and jurisdictional concerns. Clearly the large craft organizations, such as the Carpenters, Electricians, Machinists, and
others in the building trades, dominated the AFL. "This whole thing
has been so apparent to me for years," Burke declared in 1937, "that
oftentimes I have not even bothered to attend conventions."[4] He
felt that if left to its own devices, the AFL leadership would see
the necessity for industrial unionism. William Green, after all, had
been an official with the United Mine Workers and had frequently
shown concern for mass production workers. But ruthless and
unrealistic craft unions dominated the Executive Council. They
ignored production workers while refusing to relinquish their
claims or to permit other unions to recruit them. When organizations such as the paper unions did organize the mills, the crafts
soon swooped down to pick off the skilled workers. Their power
in the AFL ensured that whatever action the Federation took regarding mass organization would conform to the wishes of the
large craft unions. Burke sought to assert the isolated-communities
provision of the AFL's basic statement on jurisdiction and trade
autonomy, the Scranton Declaration of 1901. This stipulation, designed to appease the United Mine Workers, provided that in certain circumstances involving organization of workers in remote
areas, craft jurisdictions could be waived, permitting organization
along vertical lines. Burke believed that the isolated-communities
doctrine covered pulp and paper mills, which were often remote
indeed from urban centers and from access by the craft unions,
but he had little success in persuading the AFL Executive Council
of his view.[5]

3. On salaries, Burke to Earl Taylor, Jan. 3, 1941, Reel 3; on NLRB stand, AFL,
*Report of the Proceedings of the 59th Convention of the American Federation of La-
bor* (Washington, 1941), 484–86. Burke's disapproval of a statement by William
Green criticizing the United Automobile Workers in the Flint sit-down is expressed in his letter to Edward Mangan, Feb. 13, 1937, Reel 2.
4. Burke to Algernon Lee, Aug. 23, 1937, Reel 7.
5. Burke to International Executive Board and field representatives, Jan. 14,
1936, Reel 2. On the isolated-communities doctrine, see Morris, *Conflict within*

Jurisdictional problems in pulp and paper had a long and tangled history. Pulp and paper employment encompassed a wide variety of jobs some of which might conceivably be claimed by organizations other than the Pulp, Sulphite, and Paper Mill Workers. The United Brotherhood of Carpenters and Joiners, for example, could lay claims to woodcutters, sawmill workers, log haulers and handlers, and perhaps even to men employed in debarking trees and chipping wood into pulpable form, although in practice in most of the northeastern mills the Brotherhood traditionally organized these men. Inside the mills as well, jurisdictional lines became snarled, as the small groups of electricians, machinists, and firemen and oilers needed to operate and maintain the power rooms, boilers, and other machinery in some mills remained in Paper Makers' or Pulp, Sulphite, and Paper Mill Workers' locals, while elsewhere they formed separate craft locals or joined locals outside the mill. Paper union representatives regarded the crafts as parasites, skimming off the skilled workers after others had done the hard work of organizing. Craft unionists, however, resented this charge, observing that very often the original organizing bodies, especially in the mills in the South and the Pacific Northwest, comprised largely machinists and electricians, who supplied early leadership, secured a charter, and maintained a union presence until the paper unions managed to send in representatives.

The strikes of the 1920s had exacerbated the uncertain relations between the paper unions and the crafts. In the great International Paper strike, the crafts had returned to work while the Brotherhood's membership had remained on strike. Paper unionists charged that in the twenties the crafts had even recruited into their ranks defectors and strikebreakers, thus better enabling struck firms to resume production. In some mills in Canada and Maine, the paper unions had developed reasonably cordial relations with the crafts, over the years evolving informal agreements governing bargaining and membership. In the organizing upheaval of the 1930s, however, the situation grew ever more confusing, especially since hastily recruited organizers, lacking detailed knowledge of the pulp and paper industry, often signed up workers without

---

the AFL, 15–20. Morris is an excellent guide to the general jurisdictional problems discussed in this chapter. On the limitations of the term "craft unionism" in characterizing such unions as the Carpenters and Electricians, see Tomlins, "AFL Unions in the 1930s."

regard to jurisdictional proprieties. As a result, Burke complained, he was "constantly harrassed" about jurisdictional problems.[6] Burke hated the pointless and distracting disputes. He often advised field representatives to back off in confrontations with other organizations, even if Brotherhood claims might theoretically be justified. "No great harm is done . . . ," he once remarked, "if a church member gets into the wrong pew."[7] He repeatedly urged the AFL to encourage the craft unions to waive jurisdiction in industrial organizing and criticized the crafts for failing to do so. He sold employers on the virtues of organization with the paper unions—only to suffer embarrassment when the crafts intruded, detaching their members and complicating labor relations.

The most difficult jurisdictional problems occurred in areas of new growth in the 1930s. As Brotherhood organizers developed unions in the South in 1937 and 1938, they continually found the craft unions, and in particular the International Association of Machinists, complicating matters. Burke had originally hoped that the two paper unions alone could organize the International Paper Company's Southern Kraft mills. But the Machinists insisted upon exercising their jurisdictional claims and had to be included in negotiations with the company. In their preorganizing talks with J.H. Friend, vice-president of Southern Kraft, Burke and Burns had stressed the harmony and simplicity of organization under the auspices of the two paper unions. After the initial round of organizing, however, Charles Poe, the regional representative of the International Association, aggressively asserted trade autonomy, and this stance, combined with his imperious attitude toward Friend, threatened to jeopardize the multiplant bargaining arrangement that the two paper union executives sought.[8]

On the West Coast, the Brotherhood encountered more severe difficulties with the United Brotherhood of Carpenters and Joiners. The Carpenters had for years scorned mass production work-

6. Burke to Brotherhood Executive Board and field representatives, Jan. 14, 1936, Reel 2.
7. Burke to Brotherhood Executive Board and organizers, Aug. 18, 1936, Reel 2.
8. E.g., Paul Phillips to Matthew Burns, copy, Dec. 16, 1938, Reel 7. Also Burke to John Malin, Nov. 11, 1937, Reel 3; Burke to H.W. Brown, Vice-President, International Association of Machinists, Nov. 23, and Matthew Burns to Paul Phillips, Dec. 10, 1938, Reel 7.

ers in logging, sawmill, and wood-finishing operations. In 1933 and 1934, the AFL issued more than 100 federal charters to these unskilled and semiskilled workers in the Pacific Northwest. By 1935, however, the United Brotherhood had demanded and secured from the AFL Executive Council the rights to these locals, which constituted a suborganization of the Carpenters called the Sawmill and Timber Workers Union. Increasingly determined to sew up the woodworking trades on the Coast and apprehensive about the possibilities for rival unionism in the area, the notoriously strong-armed Carpenters moved aggressively to forestall competitors.[9]

The two unions had always engaged in a certain amount of jurisdictional sparring. Carpenters worked in many pulp and paper mills, often preferring the low initiation fees and modest dues structure of the Pulp, Sulphite, and Paper Mill Workers to the greater expenses involved in membership in the Carpenters, which was a benefit-oriented union. In addition, the two organizations frequently argued about the status of outside workers. Since the Pulp, Sulphite, and Paper Mill Workers held inclusive jurisdiction, covering workers employed in and around the pulp-making process, it regularly enlisted pulpwood cutters and haulers; certainly, the paper union claimed and organized workers who performed woodyard work and debarking and chipping functions at the mill site. Prior to the 1930s, the Carpenters cared little about these unskilled and often geographically isolated workers, gladly leaving them to the paper union. But in the pressurized atmosphere prevailing on the coast in the mid-1930s, the organization of woodcutting crews and woodyard and chipping workers became of deadly importance, as the aggressive Carpenters sought to enforce their sway in all wood-related operations on the coast.

On March 7, 1936, violence erupted between members of the two organizations at a Crown-Willamette logging camp at Seaside, Oregon. In the summer of 1935, Sawmill and Timber Workers' locals throughout the Pacific Northwest had gone out on strike. As the dispute began to curtail the supply of pulpwood, the Brotherhood organized loggers and woodcutters and signed contracts with Crown-Willamette. The Sawmill and Timber Workers viewed the arrangements as a direct assault on their jobs and

9. Christie, *Empire in Wood*, 287–300; Jensen, *Lumber and Labor*, 161–85ff.

threw up picket lines against members of Brotherhood Local 227 at Seaside. Never particularly fastidious in their methods, these workers, now under the jurisdiction of the Carpenters, baited and reviled the cutters, whom they considered scabs. Sawmill and Timber Workers throughout the area descended on the remote camp, jeering and fighting with the rival unionists. "Just before daybreak" on the morning of March 7, reported U.S. Conciliation Commissioner E.P. Marsh, "the assault began. Rocks were hurled at the bunk houses, autos belonging to the pulp workers were smashed and the bunkhouses [sic] were broken into and a general riot precipitated." Shots rang out, and two of the attackers lay dead.[10]

In addition to assaulting this local, the Carpenters lodged claims with the AFL Executive Council for jurisdiction over all Pulp, Sulphite, and Paper Mill Workers who handled wood in its natural state. The demand entailed turning over to the Sawmill and Timber Workers not only woodcutters but also chipping and debarking workers who toiled within or adjacent to the mills proper. While he was willing to release the cutters to the Carpenters, Burke adamantly opposed losing the pulp mill workers and protested vehemently to the AFL. He accused Carpenters' leader William Hutcheson and West Coast representative Abe Muir of "making war on our organization" and pledged to fight the huge and influential union with no holds barred. Still, certain that the Executive Council would side against the small Pulp, Sulphite, and Paper Mill Workers, in good part because it had consistently supported the industrial union position at AFL conventions and sympathized with the position of those who pressed the CIO case within the Federation, Burke retreated from a full-scale encounter. The Executive Council did turn over woodcutters and logging workers to the Carpenters, but it did eventually reaffirm the Brotherhood's traditional claims to chipping and debarking workers. The violent affair ended in an uneasy truce. The Pulp, Sulphite, and Paper Mill Workers remained embittered because of the AFL's readiness to sacrifice their interests to the powerful Carpenters. At the same time, West Coast representatives of the Carpenters warned that while they would recognize Brotherhood claims to the mill

10. E.P. Marsh to Hugh Kerwin, March 16 and 19, 1936, USCS Records, File 182-1250; "Seaside Strike Items," in file for Local 227, Brotherhood Papers, 1936, Reel 3.

woodworkers, they could not always restrain rank-and-file workers who still resented the paper union's behavior in the 1935 strike.[11]

Conflict between the Brotherhood and the craft unions in the
pulp and paper mills rarely had such drastic results. Burke usually avoided direct confrontation, knowing that the AFL would
uphold the crafts and that reopening jurisdictional questions might
prompt fresh doubts in mills long organized. Hence the president-
secretary simply sought to hold the crafts at bay, hoping to educate
their leaders in the realities of organization in mass production
industry and attempting to nudge the AFL toward an understanding of the central role played by the paper unions in strengthening the Federation and in thwarting the CIO in the basic pulp and
paper field.

The vast converting industry also proved a hotbed of jurisdictional rivalry in the 1930s. With the large number of job descriptions in the ever-changing and expanding industry, unionists could
not hope to apply neat jurisdictional definitions. Throughout the
1930s, the Pulp, Sulphite, and Paper Mill Workers collided with
unions in the printing trades, particularly the Bookbinders and
the International Printing Pressmen's and Assistants' Union, as
opportunities to expand membership in the box, container, and
novelty industries beckoned. In addition, the pulp and paper union
frequently ran into competition from the powerful International
Brotherhood of Teamsters; this organization claimed not only drivers of supply trucks but also employees engaged in warehouse and
paper-handling work around the converting shops.

In theory, of course, all jurisdictional disputes should have been
ironed out between executives of the respective unions or, failing
mutual agreement, by decisions of the AFL Executive Council. In
practice, however, even when Burke reached an accord with his
counterpart in another international union, zealous organizers in
the field continued to press their unions' claims and to challenge
the representatives of other organizations. Appeal to the AFL rarely
did much good in converted paper; the Federation showed little
interest in the industry or in the problems of "mere production
workers."[12]

11. Burke to Frank C. Barnes, Jr., Jan. 25, and to John Sherman, Jan. 11, 1936,
Reel 1; correspondence with William Green in "Seaside Strike Items," Reel 3.
12. On diversity of jobs, see Burke to Walter Trautmann, Nov. 29, 1936, Reel

From Burke on down, Pulp, Sulphite, and Paper Mill Workers' officials believed that the AFL favored its rivals in converted paper. George Berry, IPP president and a power in Democratic politics, exercised influence in the Federation far beyond the modest membership of his union. Of course, Daniel Tobin of the Teamsters exerted enormous power in the Federation, both by virtue of his political activities and because of the strategic importance of truck drivers in almost every labor dispute and organizing campaign. The Teamsters, Burke asserted, were not satisfied with the broad grant of jurisdiction contained in their charter and always augmented by AFL awards: "They want more territory, more fields to exploit, even if it means stealing away from comparatively small unions." Field representatives for the Pulp, Sulphite, and Paper Mill Workers believed that AFL state and regional functionaries automatically favored the craft unions, diverting inquiries from workers to their old comrades rather than to the less familiar representatives of the Brotherhood.[13]

In the end, however, the local situations in scores of communities determined the pattern of organization. When the CIO threat mounted early in 1937, Burke shrewdly analyzed this phenomenon. Acknowledging that the Printing Pressmen had some justified claims to folding box workers and that the Bookbinders might claim employees engaged in card making and other novelty enterprises, the president-secretary bade his representatives not to "worry too much over this matter of jurisdiction. Since the C.I.O. has entered the field," he went on, "it means that the bars have been let down . . . and it now seems to be a catch as catch can [situation] among the International Unions."[14]

This advice proved sound. In Philadelphia, neither the CIO nor

3; on printing trades competition, see Burke to Barnes, March 9, 1937, Reel 1; on Teamsters, see Burke and Matthew Burns to West Coast locals, Aug. 1, 1939, Reel 3. The quoted statement is in William Green to George Berry, April 25, 1940, AFL-CIO Records, AFL-CIO Library, Washington, Reel: National No. 10: International Printing Pressmen. For background on the AFL's policies in jurisdictional questions, see Morris, *Conflict within the AFL*; Lorwin, *The American Federation of Labor*, 49, 67, 85, 195, 339–41; Millis and Montgomery, *Organized Labor*, 204–11, 285–86, 290–300, and passim.

13. Burke to Barnes, Oct. 28, 1940, Reel 2 (quoted), and George Brooks to Burke, July 21, 1937, Reel 1.
14. Burke to Barnes, Feb. 24, 1937, Reel 1.

the Printing Pressmen contested the Brotherhood seriously in its organization of the converting industry. In Minneapolis and St. Paul, the Pulp, Sulphite, and Paper Mill Workers quickly erected a large multiplant local, nor did rivals challenge their hegemony in Milwaukee. Pittsburgh, Toledo, and southern New England also belonged to the Brotherhood. In California, however, the Pressmen waged aggressive campaigns, consistently out-organizing the pulp and paper union. New York City and nearby northern New Jersey remained a battleground, with the international quarreling with both the IPP and the Bookbinders while holding the corrugated industry for the AFL against a strong CIO challenge.[15]

In the perfervid atmosphere of the 1930s, workers shopped for union representation and found a buyer's market. To many, federal charters, with their low initiation fees and modest dues, seemed the best investment. The AFL imposed sharp restrictions on these unions, however, and most of the directly affiliated locals either were eventually absorbed by international unions or had their members parceled out among the crafts.

The Brotherhood had a considerable advantage in its competition in converted paper; its initiation fees and monthly per capita tax were among the lowest in the Federation. Certainly, these rates compared favorably with those of the Bookbinders and Printing Pressmen, whose financial and benefit structures derived from the high-wage, high-benefit printing trades. Always the skeptic, Burke believed that much of the enthusiasm for "industrial unionism" stemmed from the workers' desire to acquire inexpensive representation, and the Brotherhood was a bargain indeed. Often, he noted, workers insisted on ignoring AFL craft demarcations, forcing the paper unions to risk the wrath of other organizations, only to reassess their circumstances when they saw that higher dues of the crafts often brought higher wages. Thus, while enthusiasts in one locale clamored for industrial unionism, elsewhere skilled workers re-formed craft locals to rectify their earlier mistaken commitment to industrial unionism.[16]

15. Lambert Louy file, 1941, Reel 3; Jacob Stephan file, 1941, Reel 3; Morris Wray to Burke, Feb. 24 and Dec. 4, 1941, Reel 3; John Sherman to William Green, Feb. 21, 1941, AFL-CIO Records, Reel: National 18: Brotherhood.

16. Burke to Maurice La Belle, Dec. 2, 1937, Reel 1; Burke to Mae Pritchard, April 21, 1938, Reel 7; Burke to Barnes, Jan. 14, 1936, Reel 1; Burke to W.R. Pigman, April 8, 1941, Reel 7.

Problems associated with dues paying and membership maintenance continually plagued the Brotherhood, especially in converted paper. Without union shop contracts, and without a checkoff system, workers continually drifted in and out of the local unions. When interest diminished or a rival union threatened, local officers and international organizers ran special campaigns to re-recruit delinquent members. Back dues might be forgiven or special half-price initiations might be sold for a limited period. The more faithful members often resented these promotions, however. Cut-rate deals made *their* regular financial support seem foolish, since those who had fallen away were in no way punished for their indifference and had even been rewarded, in a sense, since they received special dues deals. Organizers had to make fine calculations on these matters, balancing the desirability of attracting delinquent workers against the resentment of more faithful customers. Throughout these elaborate campaigns and promotions, Burke clucked with disapproval, understanding the need to induce members at bargain rates but not wanting the Pulp, Sulphite, and Paper Mill Workers to become known in the labor community as a "Cheap John" union.[17]

Although at times Burke vigorously asserted jurisdictional claims, in general he viewed the problem as annoying and distracting. In the final analysis, what mattered was that workers won representation. Echoing the latitudinarianism of the nineteenth-century socialists, Burke remarked in 1941 that "it is really not so important what union the workers belong to provided that [that] union gets wages and conditions for them."[18]

Field representatives continually urged him to press the Brotherhood's claims, but Burke believed that the risks of confrontation were too great. "Our international Union is not lily white when it comes to questions of jurisdiction," he commented.[19] When a special organizer on the West Coast, appalled at the inroads of the Teamsters and Printing Pressmen on Brotherhood jurisdiction in Southern California, urged Burke to fight the rival unions aggressively, Burke was amused. "Do you mean," he asked the em-

17. Burke to Raymond Leon, Dec. 1, 1936, Reel 1; Zieger, "Limits of Militancy," 649-50.
18. Burke to Morris Wray, April 8, 1941, Reel 3.
19. Burke to John Sherman, July 24, 1941, Reel 1.

battled representative, "that I should organize a gang of mobsters and go to Los Angeles and lead my forces in a big battle against the 'goons' of the Teamsters?" Despite the inroads of the craft unions here and there and the success of the Printing Pressmen in Southern California, Burke believed that the wisest policy was to organize along the path of least resistance, avoiding sharp in-terunion conflict wherever possible. This policy, in the eyes of some of his field representatives, lacked boldness, but, Burke boasted, "I must say that our International Union for a 'little fellow' is holding its own very well."[20]

Throughout the heady days of the 1930s, the problem of juris-diction refused to disappear. In Oregon, the Carpenters attempted to outmuscle the Brotherhood. In Florida and Virginia, the paper unions fought running battles with the Electrical Workers and the Machinists, surface politeness and fraternal forms never quite concealing the mutual irritation. Even in the old, established mills, constant adjustments were necessary, as the Firemen and Oilers scrambled for every member in a declining trade. In New York and Ohio, the Bookbinders asserted claims, while in California and New York, the Printing Pressmen moved into the folding box trade. In Covington, Kentucky, the Pulp, Sulphite, and Paper Mill Workers even ran afoul of the Hod Carriers.[21]

The industry made widely diverse products and required many different operations. The Pulp, Sulphite, and Paper Mill Workers' jurisdiction began with the first bite of the saw in a remote forest in Oregon or Virginia and extended throughout the various per-mutations of the wood on its way to becoming manufactured paper and then continued as it was fashioned into an ever greater va-riety of industrial, shipping, and consumer products. As a semi-industrial union in a Federation still deeply imbued with craft

20. Burke to L. Stewart, July 3, 1941, Reel 7 (first quotation); Burke to Barnes, Feb. 24, 1937, Reel 1 (second quotation).
21. Correspondence between Burke and George Berry, July 18 and Sept. 23, 1941, AFL-CIO Records, Reel: National 8: International Printing Pressmen (IPP); Berry to Green, April 29, 1940, AFL-CIO Records, Reel: National 16: IPP; Jack Mar-golis to Burke, March 21, 1941, with copy of Bookbinders' jurisdiction statement, Brotherhood Papers, Reel 7; Burke to John B. Haggerty, Bookbinders' President, June 13, 1941, Reel 7; Fred Hook, Business Agent, Building Trades Council, Cin-cinnati, to Joe Mreschi, c. Jan. 1939, AFL-CIO Records, Reel: National 18: Brother-hood.

union traditions, the Brotherhood could not help but trespass upon the domains of the strong and influential unions in woodworking, transportation, and printing, three of the AFL's most powerful and aggressive segments. Even before the great scramble for jurisdictional rights began in earnest, Burke was both bemused by and weary of the whole unedifying process. "At the present time," he observed in April 1936, "our little International Union is involved in jurisdictional disputes with the following unions: Electricians, Boilermakers, Longshoremen, Machinists, Firemen, Engineers, Printing Pressmen and even the Paper Makers. . . . This game of jurisdiction becomes more interesting day by day."[22]

If rivalry with other AFL organizations distracted and occasionally impeded the Brotherhood, the challenge of the CIO was profoundly serious. With accelerating intensity, from the inception of the CIO as a distinct entity to the eve of the war, the Brotherhood grappled with its rival. As a union favorable to mass organization, the Pulp, Sulphite, and Paper Mill Workers supported industrial union initiatives within the AFL; Brotherhood delegates to the annual conventions voted consistently for Lewis's position.

At the 1936 convention, Burke and his cohorts opposed the expulsion of the CIO unions. Well into 1937, while deploring the split in the house of labor, Burke applauded the CIO's victories in the automobile and rubber industries and sought informal understandings with CIO representatives, many of them old comrades, regarding organizing efforts. Burke thought Lewis and his associates headstrong and unwise in disobeying the AFL, and he soon came to believe that radicals exercised undue influence within the CIO. Still, he did not originally perceive the rival labor body as a threat to the paper unions. After all, the central aim of the CIO was to organize on an industrial basis. It had work enough in the sectors neglected by the AFL, such as autos, steel, rubber, textiles, and other mass production industries. With two progressive, responsible unions in the paper industry already making steady headway organizing mass production workers, Burke saw no reason to anticipate encroachment by the CIO on his domain.

Enlistment in the CIO was not a feasible option. Burke distrusted Lewis, who he believed would prove no less ruthless and self-aggrandizing than the haughtiest craft chieftains. While the Broth-

22. Burke to Barnes, April 22, 1936, Brotherhood Papers, Reel 1.

erhood did not bar Communists, Burke and his fellow leaders feared the penetration of members of the Communist party in the affairs of the labor movement. The Pulp, Sulphite, and Paper Mill Workers had made steady progress and, as the CIO split deepened in 1936–37, were on the verge of signing and strengthening multiplant agreements throughout the pulp and paper industry. Years had been spent cultivating employers and establishing the sober and responsible nature of the union; a plunge into the fiery cauldron of the CIO at this point would impel organization in the opposite direction. Always the goal was to organize workers, to preserve and advance the union. Affiliation with the Lewis-led and communist-influenced CIO could promise only reckless adventurism.

Groups within the unions disagreed. West Coast locals passed resolution after resolution favoring CIO affiliation. As the craft unions encroached in the South, scores of kraft workers found the CIO appeal compelling. Hyman Gordon, the irrepressible president of New York City Local 107, argued constantly for cooperation, and occasionally for affiliation, with the CIO. Local unionists throughout the country peppered Burke with criticisms of the AFL. Aggressive CIO campaigns attracted workers, while the creaky AFL establishment dithered and the calculating craft unions impeded mass organization, demanding always their pound of flesh. On occasion even Burke almost succumbed. The jurisdictional squabbling endemic to the AFL "almost makes me want to join the c.i.o.," he confided to an associate in a moment of pique.[23]

But affiliation with the CIO would have required the Brotherhood to cut loose its moorings. Joining the rival federation would invite – indeed, would guarantee – ruthless attacks by the AFL craft unions. The Teamsters, Printing Pressmen, Bookbinders, and a dozen other unions would slash the Pulp, Sulphite, and Paper Mill Workers to ribbons. On the West Coast, Burke was certain, "the American Federation of Labor, through the Carpenters, could

23. Gordon and Local 107: Gordon to Burke, June 2, June 9, Burke to Gordon, June 14, 1937, Reel 3; copies of Local 107's *Paper Worker Champion*, Feb.-May 1938, Reel 5; *Report of Proceedings . . . 17th Convention* (1937), 187–89, 191. On West Coast locals: Burke to John Sherman, Feb. 6, 1936, Reel 1, and Tacoma Local 199, Resolution, Aug. 11, 1936, Reel 3. Also Shirl N. Montgomery, Local 152, Covington, Va., to Burke, May 6, 1938, Reel 4, and Burke to Raymond Richards, Nov. 26, 1937, Reel 1 (quoted).

choke us to death." As it was, the Brotherhood's consistent support for the industrial union viewpoint had brought down the wrath of the craft unions and had emphasized its isolated status in the Federation's councils.[24]

Moreover a decision to cast its lot in with the dangerous John L. Lewis would mean a complete abandonment of its previous strategy. What it had gained or was about to gain through caution and respectability it would have to win anew through aggressiveness, histrionics, and a long siege of confrontations with employers and with rival unions. In Burke's view, this sort of choice might be reasonable if employers proved unresponsive. Certainly he had no quarrel with the methods that the Auto Workers and Steel Workers Organizing Committee had had to employ. But the CIO was not magic. In the end, its purposes were the same as those of the AFL. It was unfortunate that the basic rights of organization, union recognition, and collective bargaining had to be won in bloody combat; Burke attributed the reason to the benighted attitudes of employers in many mass production industries.

Burke did contribute money and encouragement to organizations fighting the battles, at least in the CIO unions' early efforts. But the Brotherhood had fought *its* battles in the 1910s and 1920s. It had proven its steadfastness. Five years of tenacious picketing and sacrifice had paved the way for the gains of the 1930s. Why should the union opt for another, bloodier, more acrimonious round of labor conflict when it could make the steady gains and achieve the decent contracts to which its leaders' skill and experience, and its members' valor in earlier days, entitled it?

Within the confines of the AFL, Burke urged reconciliation. At the 1936 convention, Pulp, Sulphite, and Paper Mill Workers' delegates voted against the Executive Council's resolution to oust the CIO organizations. In an editorial in the *Journal*, Burke castigated both the dissidents and the AFL Executive Council for their headstrong and destructive actions. After all, the issue, he felt, was not between industrial and craft unionism per se, for the AFL had always contained industrial unions. Nor did it involve fundamental differences in trade union practice, since the CIO, despite its rhetorical lapses, acknowledged its commitment to collective bargaining and to union contracts. The only issue, really, was the

24. Burke to John Sherman, Feb. 6, 1936, Reel 1.

rather narrow matter of whether skilled workers would remain in the crafts units or would be organized in common with unskilled and semiskilled workers.

He endorsed a plan suggested by the Wisconsin State Federation of Labor that called for consultation and mutual concessions, thereby in effect publicly disassociating himself with the AFL Executive Council in its determination to oust the rebellious unions from the Federation. Always the skeptic, Burke believed that the clash of union leaders was tantamount to putting the cart before the horse. As always, the main problem of the labor movement remained the recruitment of workers and the maintenance of the organizations. "Judging by the experience that all of us have had . . . ," he reminded his fellow unionists, "none of us can be too sure that these workers will remain organized no matter what the form of organization may be."[25]

But it quickly became clear that there would be no peace. And despite the Brotherhood's conciliatory stance, CIO activitists did not permit the paper union to pursue its own course without challenge. Although the CIO initially concentrated on rubber, autos, steel, and textiles, the new federation soon began to organize promiscuously, appealing to workers of all trades and stirring enthusiasm among the unorganized. In 1936 and 1937, Brotherhood representatives reached informal understandings with some CIO functionaries, but by the end of that year the rebel body actively competed in the paper industry. Burke lugubriously observed in March, 1937, "We are between the crafts on one side and the C.I.O. on the other or between the devil and the deep blue sea."[26]

CIO plans in the paper industry developed slowly. By 1940, however, it had created local industrial unions to challenge the AFL organizations in several urban locales. UMW District 50 organizers explored the Brotherhood's locals in Virginia and farther south, while other CIO organizations began to recruit members in Southern Kraft mills. "Somehow," reported organizer Morris Wray from

25. On the stand at the AFL convention: Burke to Paul Chambers, Local 68, Oregon City, Oreg., Sept. 18, 1936, Reel 3, and to Sherman, Oct. 31, 1936, Reel 1. On support for the Wisconsin plan, Burke, Editorial, Brotherhood *Journal* 20 (Oct. 1936), 1–3, and *Report of Proceedings . . . 18th Convention* (1937), 44. I have quoted Burke, Editorial, Brotherhood *Journal* 20 (Oct. 1936), 2. Also Burke to William Koehn, Local 193, Milwaukee, Nov. 5, 1936, Reel 3.
26. Burke to Barnes, March 12, 1937, Reel 1.

Jersey City in the fall of 1938, "the c.i.o. has some magic effect with some workers." In 1941, District 50 defeated the paper unions, hampered by craft divisions, in an NLRB election involving 3,000 workers in Berlin, New Hampshire. In that same year, CIO officials claimed more than 1,000 members in the corrugated box trade alone and seemed poised to make a major commitment in the paper industry. "The CIO is everywhere these days," Burke remarked bitterly that summer, "spreading its poison and stirring up strife and confusion."[27]

The sharpest and most protracted struggles between the Pulp, Sulphite, and Paper Mill Workers and CIO elements erupted in New York City. In 1940 and 1941, Local 65 of the United Retail, Wholesale, and Department Store Employees challenged Brotherhood units in the folding, corrugated, sample card, and miscellaneous converting paper trades in the city. Brotherhood organizer Joseph Tonelli estimated that as many as 4,000 workers toiled in the corrugated segment, while the president of Local 413, which represented sample card and other specialty trades, counted at least 2,000 more in this sector. These estimates may have reflected the enthusiasm of zealous organizers, but clearly the converted paper business of New York City loomed large in importance. At stake was not only the organization of the workers in the box shops but, potentially, control over the converting industry in the entire metropolitan area. The CIO had begun to challenge Brotherhood hegemony in the large new converting mills in suburban and southern New Jersey. A victory for Local 65 in the city might well push Local 107, always an outpost of radical and CIO sympathy, into the rebel camp. And while Burke received no overt threat that boxmakers' Local 299, by 1941 the largest in the entire international union, would follow suit, it had always contained

27. For CIO activities in converting plants, see Morris Wray to Burke, Feb. 24 and July 1, 1941, Reel 3; Tonelli files for 1940 and 1941, both Reel 3; Lambert Louy file for 1941, Reel 3; John Sherman file for 1941, Reel 1; Frank Barnes file for 1940, Reel 2. District 50 and other CIO efforts in the South are revealed in a clipping from District 50 edition, *CIO News*, Aug. 11, 1941, in Matthew Burns to Arthur Huggins, Aug. 12, 1941, and Burke to Burns, Aug. 15, 1941, Reel 9. The CIO's victory in Berlin, N.H., is noted in Burke to T.A. McDonald, Jan. 12, 1941, Reel 9, and to Robert Wolf, May 15, 1941, Reel 10. Morris Wray's quoted lament occurs in his letter to Burke of Sept. 24, 1938, while Burke's mordant observation appears in his letter to J.C. Furr, July 26, 1941, Reel 7.

leftward-looking unionists impatient with AFL jurisdictional and ideological boundaries. In turn, representatives from Local 299 had been instrumental in organizing Philadelphia Local 286, the second largest in the union, and in pulling together the various converting locals in the Philadelphia-New York area into a loose district organization. Burke reported in February 1941, "We are in a big fight with the CIO in New York City right now" with a "vicious C.I.O. Local . . . [who] are doing their best to smash our organization."[28]

Both organizations pulled out all the stops. Local 65 activists visited Brotherhood unionists in their homes; Tonelli charged that these "gorillas" threatened and intimidated his people. Local 65 also called on other CIO organizations for support, gathering waterfront workers, political activists, and other industrial unionists to form flying squadrons of pickets. The CIO people correctly accused Tonelli of collusion with employers; he accused them of coercion, harassment, and union-smashing tactics. "Tonelli has signed up the second largest corrugated plant in New York," Burke reported in February 1941. "He tells me that while he was conducting negotiations . . . , a mob of two or three hundred C.I.O. members and Communists were outside . . . shouting and hissing and making threats."[29] Throughout 1940 and 1941, the young Tonelli reported on the aggressive forays of the "Communists" associated with Local 65, reports seconded by Jacob Stephan, whom Burke sent to help early in 1941. Meanwhile, local unionists in the sample card and related trades also encountered representatives of the ubiquitous Local 65, even while they beat back nagging and opportunistic jurisdictional claims by the Printing Pressmen and the Bookbinders. These conflicts sputtered inconclusively throughout 1941, occasionally flaring into minor violence. The two organizations had no final showdown, but the Brotherhood's converting locals continued to grow steadily through 1941, forming a base of power that Tonelli used to assert influence within the international union.[30]

The young organizer believed that employees in the converting

28. Tonelli to Burke, Jan. 22, 1941, Reel 3; Margolis to Burke, March 21, 1941, Reel 7; Burke to Sherman, Feb. 24, 1941, Reel 1 (quoted).
29. Burke to Sherman, Feb. 21, 1941, Reel 1.
30. Tonelli files for 1940 and 1941, both Reel 3; file for Local 413, 1941, Reel 7; Jacob Stephan file for 1941, Reel 3.

trades naturally preferred the cautious, conservative trade unionism of the Brotherhood locals and that only its resort to violence
and its willingness to employ radical fanatics accounted for Local
65's inroads. As was often the case in the AFL-CIO struggle, however, the AFL's control of local transport and service trades contributed materially to the Brotherhood locals' strength. In addition, Tonelli, Burke, and the other AFL unionists involved eagerly
sought direct cooperation with employers. Thus in March 1940
Burke told New York organizer Bernard Cianciulli to contact a
particular box company official, for "I am sure he will cooperate
in trying to get the C.I.O. out of the picture." When a key NLRB
election between Local 411 and Local 65 impended at National
Container's New York facility in the spring of 1941, Burke and
Tonelli kept in close touch with company president Samuel Kipnis. Kipnis "telephoned me while I was in Thorold [Ontario] . . . ,"
Burke noted to Tonelli, "and he urged me to come to New York
at once to help in the campaign to win the election." These tactics were necessary, all Brotherhood officials believed; fire had to
be fought with fire. "Judging by the campaign the C.I.O. has been
putting on in different parts of the country against our union . . . ,
I think that the high command has decided to do everything to
break down our union," Burke declared in the summer of 1940.
If the CIO mobilized its bullyboys and ideological fanatics, the Pulp,
Sulphite, and Paper Mill Workers had no choice but to exploit
their own strengths, namely their reputation for moderation and
realistic bargaining in the industry.[31]

    As the CIO challenge around the country mounted, Burke's as-
sessment of the rival organization became ever more biting. It was
one thing for the dissatisfied unionists to cut loose from the AFL
in order to organize the unorganized more effectively and to pro-
test against the pointless jurisdictional prohibitions that had left
thousands of mass production workers unorganized. Burke could
sympathize with this aim, but he believed it a mistake for industrial unionists to opt out of the Federation, and he feared and
distrusted John L. Lewis in his role as self-appointed savant. But
the CIO had no reason to become active in the paper industry.

---

   31. Burke to Cianciulli, March 20, 1940, Reel 2 (first quotation); Burke to
Tonelli, March 15, 1941, Reel 3 (second quotation); Burke to Sherman, Aug. 3,
1940, Reel 1 (third quotation).

The two AFL organizations with long and honorable records had made impressive gains and were rapidly making greater progress. They had consistently upheld industrial unionism and had supported the CIO unions within the AFL at considerable risk to themselves. Why, he asked CIO functionary Solomon Barkin, did the CIO "set up a dual union in the pulp and paper industry" if its central aim was to organize the unorganized? Everywhere, the president-secretary complained, "we find the c.i.o. trying to destroy clean, progressive unions like the International Brotherhood of Pulp, Sulphite, and Paper Mill Workers."

Burke could no longer believe in the originally idealistic intentions of the CIO. Its actions in the paper industry revealed it to be ruthless, destructive, and opportunistic. He grew livid at the actions of some of his old comrades in the socialist and union movements. The CIO had sent Adolph Germer, a veteran and martyred socialist and a Mine Workers' stalwart, out to the West Coast, where he had attempted to raid the Brotherhood's local in Everett, Washington. "When I learned about what that big Dutchman was trying to do . . . I would have forgotten his size and would have tried to give him a good punch in the nose," fumed the normally nonbelligerent labor leader. The CIO, Burke decided, was one of the two things that had most harmed working people and had retarded progress, the other being "the Bolshevik revolution in Russia."[32]

But if the presence of the CIO sometimes threatened Brotherhood organizations, even Burke realized that CIO actions also redounded to the benefit of the AFL union. Quite simply, the rise of the CIO stimulated workers' enthusiasm everywhere and increased organizing opportunities for almost all unions. Moreover, and most crucially, employers suddenly became most receptive to the paper unions' appeal, sometimes after years of opposition and delay.

As early as the spring of 1937, after the Flint sit-down and during the steel strike, Brotherhood organizers and officers clearly noted the trend. Burke believed that with the advent of the CIO came a whole new era of industrial relations. Employers at last

---

32. Burke's anger at the CIO and Germer was expressed to Solomon Barkin, Research Director, Textile Workers' Union of America, April 10, 15, 1941, Reel 8; his assessment of the CIO generally appears in his letter to Morris Wray, April 1, 1940, Reel 3.

grasped the message of the paper unions, namely that trade unionism stabilized conditions and benefited the company while ensuring decent standards for the workers. From Philadelphia, representative Lambert Louy reported "that the manufacturers as a
whole . . . are so scared of the c.i.o. coming in here that they are
willing to do almost anything to prevent it." Burke agreed, stressing to his organizers the need for good, firm contracts and the
importance of establishing the union permanently while employers attempted to escape the general labor unrest and the encroachment of the feared cio.[33]

Everywhere, any hint of cio activity drove employers into the
arms of the paper unions. Fulminate as Burke might about the
tactics of the rival unionists, he could not help but recognize their
enormously beneficial effect on his organization. Everywhere employers dropped their opposition to the Brotherhood, collaborated
with its representatives to direct workers into the AFL organization, cooperated in isolating cio activists, and zealously enforced
union security provisions against workers agitating on behalf of
the cio. "We can get fairly good agreement," reported a member
of Local 323 in Shelbyville, Indiana, "as the Company is really
scared of the c.i.o." In New York, NLRB election stipulations carried private agreements between the union and companies, providing for the union shop if the Pulp, Sulphite, and Paper Mill
Workers won the contest.[34]

Particularly revealing was the situation in Covington, Virginia,
where the West Virginia Pulp and Paper Company operated a mill
employing more than 1,800 workers. Ever since NRA days, the
union had sought to deal with this large manufacturer of pulp
and paper. Local 152 had been chartered in 1933 and over the years
had maintained a dues-paying membership of between 100 and
200. The company, however, conceded only what the law required.
For several years, it supported a company union in private while
denying publicly that it was doing so. It sought subtly and often
successfully to discredit local militants in the eyes of the international union's representatives. It cluttered the path toward col-

33. Lambert Louy to Burke, May 12, 1937 (quoted), Burke to Louy, May 13,
July 17, 1937, Reel 2.

34. Lillian Steffy to Burke, Jan. 25, 1941, Reel 7 (quoted); Tonelli to Burke,
March 6, 1941, Reel 3.

lective bargaining with every sort of obstacle, relying on literal interpretation of the law, voluminous correspondence, and skillfully staged delay to keep negotiations with the union in limbo. In September 1937, Local 152 won a representation election over the thinly reformed company union, compelling the company to enter into negotiations. But for four years, West Virginia Pulp and Paper bargained in form only, stalling Local 152 at every turn, although always paying great attention to legal requirements and proper decorum.[35]

Still, the serious entry of the CIO, in the form of representatives of UMW District 50, altered the situation in 1941. Clearly the company, which would vastly have preferred no union connection at all, regarded the Brotherhood as far more desirable than the CIO. Personnel Director James Towsen kept Burke minutely informed of the company's communications with the CIO representatives and with the NLRB regional office in Baltimore, to which the District 50 had appealed for a decertification election. In spring contract talks Burke found management unusually eager to cooperate and achieved good wage increases and a vacation-with-pay provision for company employees. In December, the company readily agreed to reopen negotiations and accepted an additional increase of five cents an hour. Towsen was now doubly concerned: CIO representatives had also begun agitation at the company's large plant on the Potomac in Luke, Maryland, and Piedmont, West Virginia. "I have heard nothing further from our c.i.o. friend in Baltimore re the Piedmont situation," the employee relations director reported to Burke in April, but "I will keep you advised."

The two men, equally concerned about the CIO threat, albeit for vastly different reasons, kept in close contact throughout 1941 as the AFL organization, with generous company support, beat back the challenge from District 50. Indeed, in December when Pulp, Sulphite, and Paper Mill Workers in Luke-Piedmont heard that the CIO had begun organizing the company's smaller mill in Tyronne, Pennsylvania, they quickly informed Burke so that he could

35. Ballot results in Case R-257, Oct. 11, 1937, NLRB-2, Region 5. The relationship between West Virginia Pulp and Paper, the international union, and Local 152 over the years can be traced in the yearly file for Covington Local 152. Boxes 39 and 41 of the Westvaco Papers house James Towsen's revealing files of correspondence with plant officials in Covington, with members of the Luke family, which owned the company, and with Burke.

alert management. "If the company feels there is any danger from the CIO there," advised an officer in Luke Local 36, "I don't think there would be much trouble if you would work it through the [company's] New York office."[36]

Clearly, the very existence of the CIO helped build the Brotherhood. For example, 1937, a year of vast expansion for the AFL unions, was also the first year of sustained CIO activism. Burke publicly professed to see no direct correlation, simply observing that the international union doubled its membership that year and signed union shop agreements with International Paper and with other large, formerly hostile concerns. As the president-secretary's contact with his organizers, with other union officials, and with company officials vividly revealed, however, the impact of the CIO was enormous. While its influence somewhat lessened as it became involved in political disputes and factionalism – in May 1939, Burke opined in a moment of wishful thinking that the CIO was "rapidly disintegrating"[37] – it continued throughout the prewar period to make employers receptive to the Brotherhood's appeal. Membership growth slowed in depression-ridden 1938 but shot upward again in the 1939–41 period, reaching more than 60,000 by the time of Pearl Harbor. Burke attributed this achievement to the strong reputation, skillful organizing, and responsible bargaining of his organization, but the growth also coincided with the creation of Paper, Novelty and Toy Workers' union by the CIO in 1940 and with the activity of District 50 representatives in the paper mills. Thus while Burke continued to view the CIO as an irresponsible demon, ruthlessly destroying and harassing organizations and increasingly under the sway of Communists and the megalomaniacal Lewis, in reality a substantial part of his union's success was due to the very existence of its fearsome rival.

But whatever the de facto gains accruing to the Brotherhood as a result of the CIO, the division in the labor movement remained a constant threat to the union. So far, it had done well in a dangerous game, for neither of the great national labor federations fully grasped the importance of and opportunities in the paper industry. Little imagination was needed, however, to see that a

36. Towsen to Burke, April 23, 1941, Reel 10 (first quotation); George Carpenter to Burke, Dec. 4, 1941, Reel 5 (second quotation).

37. Burke to Lambert Louy, May 2, 1939, Reel 2.

powerful CIO drive could magnify the troublesome situation in New York City a hundredfold. Nor could Brotherhood leaders dismiss the possibility that locals might be absorbed on a massive scale by the Teamsters, Carpenters, or other AFL powers in the course of their escalating conflict with the CIO.

Hence Burke actively supported moves to heal the split. "The sincere and honest members of the American Federation of Labor and the C.I.O.—and there are sincere and honest men in both camps—should speak out . . . for an end to this asinine warfare," he told a CIO unionist in 1941.[38] He continued to distrust Lewis, especially after the announcement of plans to organize the chemical and paper industries under the auspices of District 50; Lewis's daughter Kathryn was reputedly slated to play a key role. But he believed that "if John L. Lewis and Bill Hutcheson patched up their differences they could reunite the army of labor," for these two men, by 1941 both rivals of the Brotherhood, "have more power than all the millions of organized workers."[39]

Failing reconciliation, Burke hoped, the AFL might at least be purged of some of its more objectionable features. He continually sought to educate both the craft unions and the AFL national office about the realities of organizing paper workers, resisting as forcefully as he dared the constant intrusion of the Electrical Workers, Machinists, and other organizations. He believed that the everselfish crafts, and not the AFL's leadership itself, had caused most of these problems, and he hoped that George Meany, the Federation's new secretary-treasurer and "a pretty live wire," might work to resolve the jurisdictional problems. No doubt, however, Burke felt discouraged every time Green, Meany, and AFL regional representatives expressed confusion about and lack of interest in the industry; with annoying frequency they either ignored the Brotherhood or confused it with the Paper Makers.[40]

On a broader front, Burke aligned himself with the more progressive forces in the AFL. He spoke out against corruptionists within the Federation, believing the financial irregularities of some

38. Burke to Solomon Barkin, April 15, 1941, Reel 8.
39. Burke to Algernon Lee, Sept. 18, 1941, Reel 8.
40. Burke on Meany: Letter to John Sherman, July 27, 1940, Reel 1. Problems with AFL functionaries are indicated, e.g., in Burke to Paul Smith, Aug, 8, and Burke to Green, March 13, 1941, Reel 8; also Green to John Sherman, March 22, 1941, AFL-CIO Records, Reel: National 18: Brotherhood.

prominent leaders as dangerous to the labor movement as was the radicalism and irresponsibility of the cio.[41] In 1940 he publicly criticized the high salaries that the AFL convention voted for its top leadership, observing, "I still cannot understand how a union official, pulling down $20,000.00 a year and expenses can feel comfortable at a meeting of working men and women."[42] He made no friends among the top leadership when in 1939 he openly opposed the Executive Council's call for amendments in the National Labor Relations Act.

He coupled his criticism of the Federation's leadership with a call for a major effort to broaden the base and expand the activities of the Federation. He believed that the craft unions would remain a barrier to mass production organization. But he also believed that "the working people of this country are going to need the A.F. of L. . . . during the months and years ahead of us perhaps more than they have ever needed it before." Thus the doughty Burke called for a doubling of the per capita tax paid to the Federation by the international unions. The increased funding would enable the Federation to expand organizing activities, to increase service to the international unions that were bearing the brunt of the effort to organize mass production workers, and, though he did not so say publicly, to expand the power and influence of the Federation vis-à-vis the parochial and selfish craft unions.[43]

The president-secretary did not believe that his quest for reform would have much effect. "I am the president of one of the smaller unions," he remarked. "I have no great amount of influence in the labor movement."[44] He made no major commitment of his resources or those of the union to campaigns to reform and invigorate the AFL.

In reality, despite the depredations of the crafts and the challenge of the cio, these were good years for the Pulp, Sulphite, and Paper Mill Workers. The union's leaders maneuvered carefully among such powerful organizations as the Teamsters, the Carpenters, the Machinists, the Electricians, and the United Mine

41. Burke to Raymond Richards, Nov. 13, 1941, Reel 1.
42. Burke to Earl Taylor, Jan. 3, 1941, Reel 7; AFL, *Report of Proceedings . . . 59th Convention* (1940), 440.
43. AFL, *Report of Proceedings . . . 58th Convention* (1939), 388–89, 484–86, 488–90.
44. Burke to Mrs. Cele Berney, May 3, 1939, Reel 7.

Workers. The Brotherhood outstripped its rivals from the printing trades in the converting field, and by the end of 1941, despite Burke's alarms and warnings, the CIO had yet to make a major commitment in the industry.

In general, Burke was satisfied with his union's performance. Of course, given the scope and the diversity of the industry, inevitably other unions would encroach. "The C.I.O. will make some inroads," he remarked calmly; "so will the Printing Pressmen."[45] Still, all in all, the Brotherhood, which had been on the verge of extinction in 1933, was more than holding its own. Although still a small union by the standards of the craft giants in the AFL and the huge industrial unions of the CIO, it was a survivor and was building, Burke believed, a firm foundation for a strong, honorable organization which would one day take its rightful and influential place among the leaders in the American labor movement.

45. Burke to Sherman, July 24, 1941, Reel 1.

8

Change and Continuity

THE NINETEENTH CONVENTION, HELD IN TORONTO ON SEPTEM-
ber 8–12, 1941, was an occasion for reflection by Brotherhood leaders. By the eve of American entry into World War
II, the union had achieved substantial gains in membership and
improvements in the quality of contracts. At the same time, the
context in which the union operated was changing rapidly. Its Canadian members, of course, were already at war; few at the convention doubted that their American comrades would soon join
them. Indeed, in the United States, problems of mobilization for
defense and allocation of resources on a wartime footing had already begun to involve the union's officers in unaccustomed relationships with both business and government. Moreover, recent
membership gains, together with the need to function effectively
in a diverse and expanding industry, had begun to change the
union's demographic composition and internal structure.

For the time being, the old leadership remained intact. Burke
and his veteran associates were easily reelected. Still, evidence of
dramatic change was everywhere. No longer were the delegates
exclusively white males representing the pulp mills of the Northeast and Canada. Jewish, Polish, and Italian names appeared on
the roster of delegates, while three women – disproportionately few,
given the 8,400 female union members – participated as well. Men

from California, Louisiana, and Iowa mingled with representatives of the huge new urban locals. A Negro delegate, representing Local 395-A of Fernandina, Florida, addressed the body. The delegates to the 1941 convention largely rejected changes in the union's internal structure, beating back efforts to increase dues and to refine the organization's bureaucratic mechanisms. Nonetheless, the presence of global conflict coupled with the increasingly complex nature of collective bargaining in the industry suggested that the once-struggling little international union could not long remain the unsophisticated and personally oriented organization of which its veteran leaders were so proud.[1]

Since the last convention, in March 1939, the union had made substantial membership gains. In that month, its rolls had tallied about 43,000, while the figures for the first eight months of 1941 showed an average of more than 60,000, an increase of almost 40 percent. Most of this growth was concentrated in the South and in the rising converted paper locals, especially those in New York, Pennsylvania, and the Middle West. Burke counted at least 165,000 workers in the primary pulp and paper trade and as many as 200,000 more in converted paper. He believed that the union would soon reach the 100,000 mark, regardless of competition from other organizations.[2]

Burke was also proud of the union's contracts with employers. At the 1941 convention, he reported that the international organization had some 354 agreements, most of them with individual companies. Of course, the Uniform Labor Agreement covered thirty-four mills on the Pacific coast, while Local 299 in New York City had a master contract with more than 100 set-up box shops. Urban locals in Toledo, Pittsburgh, Milwaukee, Philadelphia, and elsewhere had multiplant agreements. The union had company-wide agreements with International Paper, Southern Kraft, St. Regis, Crown-Zellerbach, the Robert Gair Company, Great Northern, and other large-scale producers, which were supplemented with local agreements covering the various individual plants. But

1. Brotherhood, *Report of Proceedings . . . 19th Convention* (1941).

2. Membership figures are drawn from Auditor's Report, 1941, Reel 1. Burke's estimates are in typescript of his Report to the 19th Convention (1941), and in Brotherhood, Press Release, June 28, 1941, Reel 9. Cf. Bureau of the Census, *Historical Statistics*, pt. 1, pp. 44–45, and pt. 2, pp. 669–80.

the vast majority of Brotherhood contracts were between individual pulp, paper, or converting companies and local unions.[3] By the late 1930s the contracts—even those with the larger companies—were relatively simple documents. They defined the bargaining unit and jurisdictional stipulations and provided for some sort of union security arrangement but rarely for a checkoff. Grievance machinery typically was rudimentary and often provided CIO challengers with an opportunity for criticism. Normally the local union's officers functioned as the grievance committee; Brotherhood locals rarely developed aggressive systems providing that shop and unit stewards would be elected separately by the workers. In line with the international union's stress on good relations with its bargaining partners, grievance processes usually called for early intervention of international union representatives. This feature led to several chronic problems. On the one hand, local grievances quickly involved international organizers, who often complained at the addition to their already onerous chores. Many local unionists believed that contract negotiation and contract administration should be separate and that the practice of having local officers also serve as grievance men tended to result in the sacrifice of workers' rights for the sake of company-union harmony. Certainly, CIO activists stressed these themes in their criticisms of Brotherhood activity.[4]

Contractual fringe benefits remained rudimentary in the industry. For example, a 1941 report declared that disability and life insurance was not an attractive negotiating goal for the union, since many companies offered low-cost policies in conjunction with their in-plant safety programs. In contracts negotiated on a large scale for the first time in 1940, about 30,000 union members re-

3. Burke, Report, *Report of Proceedings . . . 19th Convention* (1941), 84.

4. These observations reflect my examination of contracts that are scattered through the correspondence and other material in the Brotherhood Papers. There is no central file of Pulp, Sulphite, and Paper Mill Workers' contracts for the period before 1942. Files of contracts from that year onward are maintained at the Labor-Management Documentation Center, Catherwood Library, New York State School of Industrial and Labor Relations, Cornell University. Weaknesses of the Brotherhood's grievance handling were stressed by local unionists I interviewed in Covington, Va., in 1980. See Zieger, "The Union Comes to Covington," 72–75.

ceived a week's paid vacation; in the 1941 negotiations this benefit was extended to an additional 20,000 workers.[5]

Wages, of course, were a fundamental issue in any contract. And
wage policy, Burke believed, was more complex than mere insis-
tence on continual increases. Since the Pulp, Sulphite, and Paper
Mill Workers represented for the most part moderately skilled and
unskilled workers, his constant aim was to increase minimum rates
and to narrow differentials between his members and the more
skilled craftsmen and machine tenders who belonged to sister
unions in the paper mills. The pursuit of a reduction in differentials represented more than a simple desire for equity. Events during the 1920s, when embattled employers pitted the Paper Makers and the crafts against the more vulnerable Pulp, Sulphite, and
Paper Mill Workers, convinced Burke that in a crisis there was
the possibility that the other unions would abandon his people.
Employers could run their mills long enough with supervisory and
skilled workers to starve out the wood handlers, chippers, debark-
ers, and general laborers represented by the Brotherhood. They
could entice the Paper Makers and the crafts back to work by agreeing to widen differentials and could then hire the more available
less-skilled workers to replace striking Brotherhood members.[6]

The narrowing of differentials thus required a subtle and cautious wage policy. Wage cutting at the onset of the depression had
affected low-skilled Paper Makers' members and in turn threatened even the high-wage machine tenders and digester cooks.
Throughout the 1930s, Burke worked with Paper Makers president Matthew Burns to raise basic scales.

The recession of 1937–39 posed a severe test to Burke's wage policy. The economic downturn staggered the paper industry. By
March 1938 it was running at only 60 percent of capacity, and
per capita paper consumption plummeted by 20 percent in one
year. The effects on workers were immediate. Employment in the
large Ohio paper and converted paper industry, for example,
dropped by one-third by the end of 1938, and take-home pay for

5. Burke, Report, *Report of Proceedings . . . 19th Convention* (1941), 85–87.
6. MacDonald, "Pulp and Paper," 109–13; typescript of Burke, Report to 18th
Convention, March 1939, Reel 1, and to 19th Convention, Sept. 8, 1941, Reel 9.
For a critical view of Burke's attitudes toward wage issues in the 1920s, see Cernek, "Beyond the Return to Normalcy," 76–126.

those employed fell from an average of twenty-six dollars a week to less than twenty-three dollars. Hourly rates declined by 10 percent. Improvement in 1939 and 1940 occurred only slowly.[7]

These conditions brought the experiences of the early 1920s forcefully to mind. In those days, the Pulp, Sulphite, and Paper Mill Workers had followed the lead of the Paper Makers and had pursued a "no backward step" policy. This course of action had eventually brought disastrous strikes that had devastated the union. In some cases, Burke believed, the Paper Makers had abandoned their brother unionists, accepting employers' offers of favorable wages for skilled workers, thus permitting paper companies to run their mills and to outlast the striking unskilled Brotherhood members. It had, of course, taken the union more than fifteen years to repair the damage.

Thus in 1938 when employers broached the subject of reopening contracts to gain wage concessions, they found Burke sympathetic. Convinced that the downturn was only temporary and that employers would be tempted to undercut the union if it did not grant wage relief, Burke overrode the objections of local unionists to renegotiation. By accepting the 5 and 10 percent reductions, the union could demonstrate its maturity, discipline, and good faith. Better, Burke believed, to hedge wage reduction with conditions and limitations reached through negotiation than to encourage employers to begin pitting worker against worker and to launch round after round of wage cutting.[8]

Even so, local unionists rarely appreciated these long-range calculations. Southerners proved particularly recalcitrant, rejecting the first union-management initiative and subjecting Southern Kraft officials to hostile questioning before eventually agreeing to temporary wage cuts. Employers, of course, were more understanding. The general manager of International Paper's book and bond

---

7. Burke to George H. Adams, March 25, 1938, Reel 4. Guthrie, *Economics of Pulp and Paper,* chart, 66; Boothe and Arnold, *Earnings . . . in Ohio Industries,* 386–403; Maclauren, "Wages and Profits," 208–14.

8. Maclauren, "Wages and Profits," 208–10; files for Southern Kraft Company and International Paper, Book and Bond Division, 1938, Reel 7; Burke to Jacob Stephan, Feb. 4, Burke to George Brooks, June 10, and to Raymond Richards, Oct. 22, 1938, all on Reel 1; L.H. Gaetz, General Manager, Folding Box and Container Division, Robert Gair Co., to plant managers, Burke, and Matthew Burns, Jan. 19, 1938, Reel 7.

division thanked the president-secretary "for the cooperation and sound judgment . . . you used in arriving at the acceptance of the 5 percent decrease in wages." The head of the Robert Gair Company's folding box and container division worried lest Burke's statesmanlike action cause disaffected unionists to withhold their dues. Since in the wage negotiations the paper unions had "done their job with us in . . . an admirable manner," he directed company officials to make "a real attempt . . . to assist them in securing a complete membership of fully paid up members."[9]

Throughout this baleful period of wage cutting, Burke counseled cooperation, restraint, and patience. "The hardest test that any trade union official has to meet," he told delegates to the 1939 convention, "is to lead those he represents in a retreat." But the alternative, he believed, was to plunge into unwinnable strikes and to revive the likelihood that the less-skilled workers would be victimized again by the companies and by the more highly skilled employees. The union could not change the economic facts of life; there came a time when there was no choice: "the army of labor [had] to execute an orderly retreat."[10]

Even when stepped-up defense production began to bring the national recession to an end, Burke continued his moderate approach to wage policy. As pressure from the locals for rescinding the 1938 wage cuts and for rebuilding wage rates increased, he urged restraint. Wage increases seemed to Burke almost as dangerous as wage cuts. Employers were always alert to means of increasing differentials that might bring opportunities to undercut the unions, to tighten plant discipline, to promote productivity, and to redefine job classifications. By 1941, Burke felt, startling CIO gains in basic, durable goods industries such as autos, steel, and national defense had given rise to unrealistic expectations among pulp and paper workers, firmly located as they were in the

---

9. Opposition by southern unionists: Shirl N. Montgomery, Covington, Va., Local 152, to Burke, Oct. 6, 1938, Reel 4; Minutes, Panama City Conference (July 8–9); Minutes, Mobile Conference (Aug. 15); Joint [local union] Committee to J.H. Friend, Vice-President, Southern Kraft Company, July 9, 1938, all on Reel 7. Company praise for international union leadership: C.E. Youngchild to Burke, Aug. 31, 1938, and L.H. Gaetz to plant managers, Burke, and Burns, Jan. 19, 1938, Reel 7.

10. Typescript of Burke, Report to 18th Convention, March 1939, Reel 1, and typescript of Report to 19th Convention, Sept. 8, 1941, Reel 9.

nondurable goods category with its historically lower wages. "At present time," he confided to employer Robert Wolfe in May 1941, "the workers turn up their collective noses at any increase less than ten cents an hour."[11]

Despite his long-range concern regarding wage structures, Burke sometimes seemed out of touch with rank-and-file paper workers. His warnings about the dangers of overexpectation and the need for orderly advance could sound mean-spirited and complacent to workers earning forty-five and fifty cents an hour. "The members of many of our newly organized locals," the president-secretary commented, "seem to think that all there is to the trade union movement is getting a wage increase." On one occasion he added, "Oftentimes, we are able to build better Local Unions when they have to struggle a little for all they get." A Cleveland box worker replied to one of Burke's lectures about the dangers of militant wage demands. "The employees here are making hardly enough to keep going. They have to work hard for the few cents [they get]. . . . How," she asked, "would you like to work all week and get an $18.00 pay?"[12]

By the spring of 1941, the union was able to resume its upward revision of scales. Brotherhood bargainers insisted on negotiating in terms of flat increases expressed in cents per hour rather than acceding to employers' desire to express wage gains in percentages. The union's method weighted increases in favor of lower-paid workers, in keeping with the program of narrowing differentials. Conferences with Southern Kraft and Pacific Coast employers led to seven-cent and ten-cent additions to base wages. Minima now stood at 54 cents for whites and 50 cents for "colored" workers in the South and at 75 cents for men and 62.5 cents for women on the coast. Pressure from the ranks forced the reopening of these contracts in the fall, bringing additional increases in a booming wartime economy.[13]

---

11. Burke to W. Rupert Maclauren, Oct. 18, 1939, Reel 7; to Frank C. Barnes, Jr., Nov. 13, 1940, Reel 2; to Crowell Pierce, Sept. 27, 1941, Reel 7; Burke to Robert Wolf, May 15, 1941, Reel 10 (quoted).

12. Burke to Barnes, Nov. 13, 1940, Reel 2; Olga Susteric to Burke, May 15 and 20, and Burke to Susteric, May 17, 1941, Reel 4.

13. Burke, Report, *Report of Proceedings . . . 19th Convention* (1941), 87. By the end of the year, contract improvements had raised southern minima to 54 cents for Negro labor and 58 cents for white workers. Burke to Raymond Richards,

By 1941, the Pulp, Sulphite, and Paper Mill Workers had indeed increased wages for members and had narrowed differentials. In 1934 common labor in New England mills was paid at rates 50 percent less than those earned by machine tenders, but by 1941 rates had risen to 57.6 percent. Workers in the lake and middle Atlantic states experienced similar gains, while on the coast common labor rates rose from a 45.9 cent average in 1934 to more than 75 cents in 1941, placing common rates at more than 60 percent of those paid to tenders, in contrast with the 49.5 percent prevailing in 1934.[14]

Gains in the southern mills were dramatic, too. Differentials there had been unusually great, as newly opened mills paid experienced machine tenders and digester cooks large premiums to relocate from the North. At the same time, of course, the notoriously low scales for common labor in the South widened the gaps. Thus in 1934 machine tenders were averaging $1.07 per hour, while rates for common labor stood at just above thirty cents, a wage rate only 28.7 percent that of the tenders. But by 1941 contracts negotiated by the paper unions helped to raise common labor rates to fifty-four cents (fifty cents for "colored" workers) and to reduce the differential to about 40 percent. When Burke proclaimed, as he frequently did, that the Brotherhood had made outstanding gains for common labor everywhere and that his organization was spearheading the cause of ordinary workers in the South, he referred to this important shift in the internal structure of wages within the industry and to the increasingly advantageous position enjoyed by pulp workers vis-à-vis their more skilled co-workers.[15]

The economic uncertainty of the late 1930s and early 1940s, with its contradictory pressures on wages and negotiations, coupled with

---

Dec. 12, 1941, Reel 1. The same letter also comments on the flat and percentage methods of adjusting wages.

14. MacDonald, "Pulp and Paper," table, 110–11.

15. Ibid.; Burke, Report, *Report of Proceedings . . . 19th Convention* (1941), 87; Burke to Brooks, June 17, 1940, Reel 1; to Lewis Williams, Local 395-A, Fernandina, Fla., n.d. but c. 1941, Reel 4. Also Burke's statements at NRA code hearings, 1933 and 1934: e.g., Hearing 405-1-04 (Paper and Pulp Industry—Proposed Revisions), Sept. 14, 1933, NRA Records, Records Maintained by Library Unit, Transcripts of Hearings; and Brotherhood representations in connection with establishing minima relating to the provisions of the Walsh-Healy Act of 1936, as revealed in correspondence with Matthew Burns and Carl Whitney, counsel for American Pulp and Paper Association, March 18–Oct. 28, 1938, Reel 7.

the increasingly vigorous CIO challenge, provided ample work for union functionaries. Travel schedules expanded, and bargaining sessions became increasingly complex. Further crowding the agenda was the need to deal with major governmental initiatives in the field of labor conditions and industrial relations. The passage in 1938 of the Fair Labor Standards Act, the ongoing work of the committees that were established to determine wage and hour provisions, and the creation in 1941 of the Office of Production Management to coordinate defense activities all entangled the Brotherhood in the federal bureaucracy to a degree reminiscent of the days of the NRA.

In reality, the actual issues with which these federal bodies dealt directly affected the union and its members only marginally: minimum wages in even the most depressed sectors of the converted paper trades rarely fell below the twenty-five-cent minimum initially required by the Fair Labor Standards Act, and even the forty-cent minimum established in the fall of 1940 for most of the industry affected only about 15 percent of converted paper workers and hardly any in primary pulp and paper. Nor did paper industry representatives and spokesmen play a major role in the affairs of the Office of Production Management; the government concentrated on industries more directly involved in national defense. Union leaders regarded *representation* on the various boards and agencies as crucial, however, for Burke soon found that the federal bureaucracy had created an additional arena in which his organization had to demonstrate its right to speak for pulp and paper workers and in which the organization had to meet the challenge of the craft unions and of the CIO.[16]

16. Phillips L. Garman, Research Director, International Printing Pressmen and Assistants' Union, to Boris Shishkin, March 5, 1940; U.S. Department of Labor, Wage and Hour Division, Press Release, July 16, 1940; Industry Committee 14, "Report and Recommendations" (draft), in Boris Shishkin to Burton E. Oppenheim, Director, Industry Committee Branch, Wage and Hour Division, U.S. Department of Labor, Nov. 20, 1940; Wage and Hour Division, Press Release, June 6, 1941: all in AFL Papers, State Historical Society of Wisconsin, Series 4: Industry Reference Files, Box 97. Also, Sullivan, Statement, Department of Labor Public Hearing on Wage Determination in the Converted Paper Industry, Sept. 1940, Brotherhood *Journal* 24 (Nov.-Dec. 1940), 5-7, and Burke to Gustav Peck, Assistant Director, Hearings Branch, Wages and Hours Division, July 18, 1941, Reel 8. And (concerning the Office of Production Management, or OPM), Matthew Burns, AFL Labor Representative, Labor Division, Priorities Branch,

The further intrusion of government into labor matters was a mixed blessing, insofar as Burke was concerned. Although at times he enthusiastically acknowledged the role of the NRA and the National Labor Relations Act in building his union, he remained skeptical about the overall efficacy of federal action, particularly when it sought to exert a direct impact on the economy. He criticized the New Deal for waste and extravagance, arguing in 1941 that "if a part of the vast sums of money that have been spent on relief work by our government were expended in helping and encouraging private industry to employ more and more workers, I think that much greater progress could be made in finding a . . . solution to the problem of unemployment." He considered committees and boards established to advise on or to implement government policies, such as those that functioned during the NRA period, inherently biased against labor. Although he was a staunch antitotalitarian who spoke out early and often against the rise of fascism in Europe, Burke nonetheless viewed the movement of the economy toward a wartime footing with apprehension. It brought, he said, a false prosperity, an inflated price structure, and the illusion of full employment, encouraging workers to make unrealistic demands and furthering the mistaken impression that economic justice had been achived.[17]

Moreover, he feared that the growing involvement of the government in the economy would undermine republican virtue and civil liberties. Burke believed that FDR's victory in 1936, while obviously benefiting the labor movement, came close to awarding the president dictatorial powers. As much as he sympathized with the democratic countries in their struggle against the Nazis, he refrained from active participation in such groups as the Committee to Defend America by Aiding the Allies. "I believe," he told a supporter of the committee, "that President Roosevelt's lend-lease bill should not be passed without amendments that will safeguard our democracy here at home." The lend-lease bill gave Roosevelt too much power. However much trade union officials might sympathize with the Allies, "it is very important that . . . [they] keep

OPM, to Burke, Aug. 25 and 28, and Burke to Sidney Hillman, Sept. 3, 1941, Reel 9.

17. Burke to E.W. Kiefer, Feb. 4, 1941, Reel 10 (quoted); Burke to W. Rupert Maclauren, Oct. 18, 1939, Reel 7; Burke to Barnes, Nov. 13, 1940, Reel 2.

their heads in this crisis," for the specter of events in 1917 and 1918—the imprisonment of the sainted Debs, the raids on socialist facilities, the jailings and deportations—loomed large whenever the government gained emergency powers.[18]

Burke also counseled against direct political commitments. His personal politics, of course, were of the third-party variety. In 1914 he had been Socialist candidate for governor of New Hampshire. He had supported Robert M. La Follette for the presidency in 1924 and in the election of 1940 voted for Norman Thomas, just as he had cast his ballots for Debs in earlier elections. He believed that there was little difference between Republicans and Democrats and scorned labor politicians such as Daniel Tobin and George Berry who rushed to join the Roosevelt bandwagon. Workers, he believed, would vote as they pleased, and it was foolish for trade unionists to risk antagonizing them by pressing a political line.

Old comrades pressed him to support the creation of a labor party. He replied that, since workers would not support their own organizations without union security devices, it was unrealistic to expect them to provide financial backing to a political party. However beneficial its programs might be to workers, they would inevitably perceive it as abstract and remote from their daily concerns. Who, he asked, would collect from the workers the dues that were to go into the labor party's coffers?[19]

In short, the president-secretary remained wearily skeptical of both government intervention in the economy and political activity on the part of organized labor. He never developed his views into an explicit, full-scale critique of the emerging corporate capitalist national defense state. Certainly, he eagerly grasped opportunities that government initiatives provided to expand organization. Still, Burke never celebrated the disguised social democracy of the Democratic liberal-labor nexus embraced by so many of his fellow labor chieftains. Organized labor, in prospering under

18. Burke to Irwin Jaffe, Jan. 22, 1941, Reel 9 (quoted); Burke to Robert B. Wolf, Nov. 8, 1936, Reel 4.

19. On the labor party: Burke to Raymond Richards, July 24, 1936, Reel 1, and to John Bayha, Aug. 28, 1940, Reel 4; on politics generally, Burke to George Brooks, Jan. 6, 1938, Reel 1, to Morris Wray, Dec. 7, 1939, Reel 3, and to Wray, Jan. 17, 1941, Reel 3; critique of politically oriented labor leaders: Burke to Burns, Aug. 7, 1941, Reel 9.

a Democratic administration committed to welfarism and global interventionism, he believed, could easily find that it had lost its independence and had become but a cog in a giant political apparatus. He continued to view his union as an outpost of independence in an increasingly bureaucratic society, as a simple, direct expression of the workers' interest, with need of neither political paraphernalia nor bureaucratic elaboration.

By 1941, the enormous changes in the environment in which the Brotherhood operated had begun to transform its character. While it remained bureaucratically primitive, by the eve of American entry into World War II it contained 337 local unions and enjoyed an annual cash flow of nearly $500,000. In 1941, Burke employed some forty-six organizers, thirteen more than in the previous year and twenty-one more than in 1938. The 1937 convention had added a seventh vice-president, and groups of local unions in particular locales had begun tentatively to add full-time functionaries to their staffs and to create district and product suborganizations. Still, almost a decade of growth under the New Deal found the Brotherhood's basic structure little changed from the period before its fifteenfold expansion.[20]

The most significant underlying change in the organization's character was the expanding importance of urban converted paper locals in its membership. The rise of converted paper, hardly represented at all in the union of 1933, sent forth new leaders just now beginning to be influential in the international union. Local 286 in Philadelphia and various New York City locals spearheaded this development. Led by energetic young unionist Henry Segal, Local 286 boasted more than 1,200 members by mid-1941, having organized more than forty set-up box shops and having forged a multiplant agreement with area employers. Extending its territory into folding box, boxboard, and envelope shops as well, the local sent its people into the rapidly expanding converting shops throughout southeastern Pennsylvania and nearby New Jersey.[21]

As always, New York City led the way in expanding organization of converting workers. And as always, Burke and other old-

20. Index, Brotherhood Papers, Cornell University; Auditor's Reports and officers' monthly reports, 1941, Reel 1.

21. File for Philadelphia Local 286, 1941, Reel 6; Brotherhood, *Report of Proceedings . . . 19th Convention* (1941), 104–105.

time Brotherhood leaders viewed the rise of the city locals as a mixed blessing. The two oldest locals—Hyman Gordon's radically inclined Local 107 and A.N. Weinberg's busy Local 299, which had 2,000 members—urged Burke to use them as nuclei for further expansion, but he searched for other leadership. He distrusted the radicalism of Local 107 and believed the Box Makers to be erratic and unreliable. Moreover, Burke and other veteran Brotherhood representatives felt that the sophisticated, often disputatious members and officers of these locals held them in disdain. "I presume," the president-secretary remarked in calling an organizing success to the attention of one of Local 299's most articulate activists, "that you will notice that our International officers are not the dumbbells that some of the 'wise guys' among the paper box workers in New York have insinuated in past years." He defended his veteran organizers to Hyman Gordon, who was often impatient with the Brotherhood's caution and conservatism. "The men who represent our International Union, . . . may not be as brilliant as some others," he admitted, but they "are men who have worked at the trade."[22]

Bypassing both of the established locals, Burke instead made Joseph Tonelli of Bronx Local 234 his key man in New York. The young activist had come to Burke's attention after he led a successful strike in 1936 and established his local as a 100 percent dues-paying organization. In 1939, Tonelli received a "permanent" organizing position and began his role as point man in the Brotherhood's struggle with the CIO in and around New York City. "I believe that we shall be able to develop Tonelli into some kind of leader," Burke observed in 1939.[23]

Although he came from rural Pennsylvania, Tonelli had adapted to life in New York, and he was well suited to the position given

22. Burke to Harriet Wray, Jan. 22, 1938, Reel 5; Burke to Hyman Gordon, May 22, 1939, Reel 5. Burke's letter to Gordon, July 6, 1935, Reel 3, provides background on the history of unionism in New York's converted paper industry. My interview with George W. Brooks of Cornell University, Feb. 7, 1979, also helped untangle the skein.

23. Stephan's remarks, based on his experiences with Tonelli's Local 234 in 1936, at 1944 convention, in *Report of Proceedings . . . 20th Convention* (1944), 268; Sullivan to Burke, Dec. 12, 1936, Reel 2; Sullivan in Brotherhood *Journal* 23 (July-Aug. 1939), 4–5; Burke to Sullivan, Dec. 29, 1939, Reel 1.

him. He ran his local unobtrusively. Moreover, he felt that in the battle against the CIO, cooperation with employers was more important than establishing proletarian credentials. Bypassing the older locals, he and his able organizing partner, Bernard Cianciulli, channeled new recruits into newly created Locals 413 and 381, preventing expansion by both of New York's pioneering converting locals while building up a multiplant membership that looked to Tonelli for leadership.[24]

Burke and his colleagues appreciated Tonelli's success and rewarded him with rapid advancement. New York City's large converting industry membership, with Tonelli in control, could now safely be added to the Brotherhood without fear of expanding the influence of Local 107's ambitious leftists or of Local 299's independent-minded activists. Representatives of these locals resented Tonelli's sudden rise. Gordon in particular charged that the Bronx unionist was creating bogus locals, diverting workers from their proper niches in Locals 107 and 299, and working with stooges, foremen, and employers to strengthen Local 381.[25] But no one in power paid attention. The union's traditions, as well as its day-to-day campaigns, seemed safe in Tonelli's hands.

At the Toronto convention, the forces that were changing the Brotherhood were clearly in evidence. Converted paper delegates, most of them representing urban locals, composed about a quarter of the individuals present. The meeting place, Toronto, reminded the convention of the presence of war and of the likelihood that the union would be further absorbed into the labyrinth of gov-

24. On the occasion of Tonelli's election as ninth vice-president in 1944, Burke said, "You notice he is a pretty snappy dresser—double-breasted suit and all that stuff there. Looks like Broadway" (*Report of Proceedings . . . 20th Convention* [1944], 268). Tonelli was the first executive board member elected from a converted paper local. For Tonelli's biography, *Report of Proceedings . . . 28th Convention* (1968), 3. Tonelli's file for 1940 and for 1941, both Reel 3, detail his activities in New York City. See, e.g., Tonelli to Burke, Feb. 27 and Aug. 12, 1940, Reel 3.

25. Burke often praised and defended Gordon and his Local 107 activists, believing them to be good trade unionists and dedicated fighters. See his letters to Cianciulli, March 9 and Dec. 20, 1940. But he did not regard them as the most effective or controlable forces with which to spearhead expansion in the key New York City converted paper field. Gordon's many and sharply worded warnings about Tonelli's methods are seen in his correspondence with Burke, Local 107 file, 1940, Reel 5.

ernmental and trade union bureaucracy. True, the international union still retained much of its old simplicity. Its affairs were still run directly from the offices of the president-secretary in New York State. It employed no staff of lawyers, statisticians, economists, or regional directors. On the other hand, many of the delegates believed that the union had to streamline its procedures, beef up its treasury, and set aside the quaint parochialism that continued to characterize its operations.

Delegates were confronted with no proposals for drastic overhaul of the constitution, first adopted in 1906 and changed only in minor details since. At the same time, proponents of change sought to streamline the union and to bolster the resources that its leaders could command. There was lengthy debate of motions to increase dues, to make the checkoff a prime bargaining target, and to bar Communists and subversives from membership in the union. In each case, the delegates voted down proposals favoring these initiatives, but the disputation caused advocates of the union's old traditions to wonder just how long the growing organization could maintain its view of itself as a simple, voluntaristic trade union.

At the convention, Burke hoped that the delegates might be persuaded to support a dues increase. By late 1941, the union had a strike fund of nearly half a million dollars, and few demands had been placed upon it. Still, the president-secretary believed that if the union was to accelerate its organizational efforts and to battle the cio effectively, the international office needed more money. Burke believed that the converting locals, many of which had large numbers of female workers, chronically cost the international office more to service than they ever returned in per capita tax; large numbers of the workers newly recruited in the cities were women who paid a per capita tax of only forty cents a month to the international union. As the Brotherhood moved ever deeper into the converting field, it would face increasingly more costly organizing prospects. Where once a differential in dues for women made sense, wage rates were now sufficiently high to justify equality of support for the union. Since the union had about 8,400 female members, the international could expect to increase its dues collection by slightly more than $20,000 a year if the motion to increase women's dues gained approval. This amount would represent about 5 percent of its yearly income, a sum sufficient to permit the organization to put three additional representatives in the field.[26]

Although Burke urged adoption of the increase, he left the issue up to the resolutions committee and to the convention. Locals with large numbers of female workers united in opposition. Some male delegates believed that it would be unchivalrous for the male-dominated convention to increase women's dues. Moreover, a dues increase would play into the hands of the CIO and antiunion employers. Even though women had shared in recent wage increases, their renumeration and conditions of employment continued to compare unfavorably with those of men. Some delegates believed that a dues increase for women now would only pave the way for a general increase at the next convention; others felt that the effort to raise dues was an indirect effort to hamstring the new urban locals' organizing activities and thus to favor the union's traditional power centers. At any rate, a Louisiana delegate declared, the proposed increase would surely complicate affairs for the officers of Local 362, to which he belonged, because the women of Bogalusa "are always looking for an excuse to drop out of the union." Burke's dues increase did not pass. Both the resolutions committee and the convention voted against it. Although he deplored this result, the president-secretary did find one note of comfort: for the first time since the 1935 convention, dissident delegates did not move to cut the per capita tax.[27]

The veteran leader was less certain about another dues-related initiative, the proposal to have the international union proclaim the checkoff as a prime bargaining goal. He was, of course, ambivalent about the whole subject of union security, deploring the need for union maintenance agreements on the one hand while energetically cooperating with employers to enforce them, espe-

---

26. Burke on dues increase: to Walter Trautmann, July 19, 1941, Reel 3. On strikes, undated form, c. late 1940–early 1941, Reel 9 (file for U.S. House, Naval Affairs Committee). The report indicated that in 1940 the Brotherhood paid out $5,647.81 in strike benefits and that 1,000 workers in all were involved. Burke calculated that it cost the union between $5,000 and $7,000 a year to keep an organizer in the field (Burke to John Sherman, July 24, 1941, Reel 1, and to Andrew Woessner, Dec. 22, 1941, Reel 6). The number of female members is given in Burke, Report, *Report of Proceedings . . . 19th Convention* (1941), 83.

27. Brotherhood, *Report of Proceedings . . . 19th Convention* (1941), 132–41; I have quoted from p. 135.

cially against CIO enthusiasts, on the other. Once the need for the union shop provisions was accepted, the checkoff seemed a natural concomitant. To urge the one while resisting the other was to choke on the gnat after swallowing the camel. Yet Matt Burns, a former president of the Paper Makers who was now helping the Brotherhood as a special organizer, regarded the checkoff as pernicious. "Over the long pull," he said, "the check-off is demoralizing so far as trade unionism is concerned. The check-off will do to trade unionism what Emperor Constantine's conversion to Christianity did to real Christianity." Burke was less certain. The United Auto Workers had received the checkoff from Ford, and the United Mine Workers were about to get it in their paper mill in Berlin, New Hampshire. Perhaps the paper unions had no choice.[28]

To some Brotherhood unionists, the checkoff was a simple matter. Henry Segal, a leader of Philadelphia's large Local 286, for example, regarded it as a necessity in the complex, multiplant converting field. It was a simple matter of accounting for his local, which represented more than forty plants and many contracts. Union shop provisions, Segal told the delegates, were difficult and awkward to enforce. Often employers applied them selectively and used the veiled threat of lax enforcement as a negotiating weapon. Workers, Segal believed, possibly differed in the modern age from their counterparts in the union's distant early days. Urban box workers had many distractions, many demands on their financial resources. Quite frankly, the rising young unionist informed his colleagues, workers now had to be compelled to remain in good standing. And of course, the "rival organization"—the CIO— felt no embarrassment about securing the most ironclad checkoff provisions possible.[29]

But many delegates believed the checkoff high-handed and cal-

28. Burke's generally positive attitude toward the checkoff is apparent in his letters to Joseph Hora, Feb. 28, 1936, Reel 3, and to Morris Wray, July 7, 1941, Reel 3, but he did not consider it a priority item. Burns's assessment is given in his letter to Burke, June 21, 1941, Reel 3.

29. The debate is in Brotherhood *Report of Proceedings . . . 19th Convention* (1941), 104–105; see also Segal to Burke, Oct. 2, 1941, Reel 6: "If we did not get this check off system" in recent contract negotiations, "the men would never have paid, for some of them the following day after the strike wanted to know what [had] happened."

lous. Union shop provisions were indeed necessary, but at least the union still collected its own monies. The automatic deduction and collection of dues by the employer, on the other hand, put him and the union on the same side insofar as the rank-and-file member was concerned. New York City Local 299 delegates, who had to monitor union security involving workers at far more than 100 box shops, vigorously opposed Segal's motion. The check-off, they admitted, might be more efficient, but forty years of struggle to form their union made them reject the very idea of relying on the bosses to act as a collecting agency. The convention supported the resolutions committee's recommendation, which rejected Segal's motion.[30]

Thirty months had passed since the 1939 gathering. Now the delegates amended the constitution, permanently increasing the time between conventions from two to three years. The change was supported by the executive board as a means of saving money; it also could not help but increase the autonomy of the executive officers, rendering the central headquarters less answerable to the rank and file and to the local unions. Still, the delegates had defeated the dues increase and had turned their backs on the checkoff as a primary bargaining goal, indicating that many in the organization continued to view the union as a relatively simple and direct voice of the ordinary pulp and paper worker. Nor did resolutions aimed at barring Communists and "subversives" from membership make headway;[31] however enmeshed workers were in the politics and economics of global conflict, there seemed to be little interest in gratuitous scare-mongering.

The Toronto convention occurred at a time of rapid change for the union. The membership figures, Burke's optimistic reports on contracts, the gains reported even for chronically ill paid converting workers all bespoke a bright future. Certainly veterans of the organization blinked with incredulity at the sheer size of the convention. Delegates from more than 200 local unions represented workers who made goods that, in many cases, had not even existed at the time of the pulp workers' initial gathering in 1906. At the same time, however, the convention met in a country at war, and no one could be certain of the consequences to the union

30. Brotherhood, *Report of Proceedings . . . 19th Convention* (1941), 104–106.
31. Segal also sponsored an antisubversive resolution. Ibid., 106–107.

of the inevitable American participation in the conflict. Meanwhile, the CIO threat was on everyone's mind, and the delegates heard a lengthy analysis and criticism by Sullivan of the disruptive and obstructive role of the AFL craft unions in the pulp and paper industry.

Whatever the organization's new-found success, many questions remained. Thinly disguised behind the pride of accomplishment of veteran unionists were apprehensions that the new recruits from the converting sector would change the union's basic character. Others worried about the Brotherhood's vast geographical diffusion and the extent to which locals in remote sections would remain loyal to an organization whose roots remained so firmly planted in the East. The more traditional unionists were uneasy over the extent to which their organization, with its mixture of socialist politics and voluntarism, could continue to flourish in a new political climate, one already being intensified by the demands of world conflict.

A handful of union dissidents, joined by increasingly active CIO elements, raised other questions. How, they asked, could a union so dependent on cooperation with employers truly represent its members? Even as the 1941 convention met, paper workers in the Middle West and upper South were beginning to look for more aggressive representation. True, by 1941 the CIO had made only scant inroads into Pulp, Sulphite, and Paper Mill Workers' jurisdiction, but having now largely completed its organization of auto, steel, rubber, and other mass production industries, would it not turn its attention to a modern industry employing perhaps a half million workers? And would not full employment in the burgeoning mills make the cautious, collaborative wage policies and the diffident grievance handling practiced by the Pulp, Sulphite, and Paper Mill Workers seem archaic? Far from being impressed with the Brotherhood's growth in the 1930s, these dissidents believed that the organization had realized only a fraction of the potential that the New Deal years had offered.[32]

32. CIO Staff Memo, Feb. 3, 1944, "Paper Workers of America," Reel: CIO National 2, AFL-CIO Records; clipping, dated Oct. 4, 1941, from CIO source, and John Fleshman to Burke, September 13, 1941, both in file for Covington, Va., Local 152, 1941, Reel 4. The observations in this paragraph regarding criticism of the Brotherhood also drew upon interviews with Nicholas Vrataric (May 9, 1980), Clarence Farmer (June 4, 1980), and David Scott and Leonard Jones, Sr. (June 4,

Such questions percolated beneath the surface as the 1941 convention came to a close. Yet for John Burke and his close associates, they remained rather abstract. No sooner had the delegates departed than rank-and-file workers throughout the Southern Kraft bargaining structure, joined by Brotherhood locals in the West Virginia Pulp and Paper Company's mills, were clamoring for a reopening of contracts negotiated earlier that year. Organizers reported that CIO and AFL craft incursions in Brotherhood jurisdictions were increasing. New vistas beckoned: "I have decided that the time is ripe to launch an intensive organizing drive in New Jersey," Burke announced in November.[33] He had little time for long-range planning; the future would have to look after itself.

1980). Vrataric was an organizer and officer in the various organizations established through the CIO in the paper industry in the 1940s and 1950s; he was at the time of our conversation vice-president of the United Paperworkers International Union. Farmer, Scott, and Jones were active local unionists in Covington, Virginia, where in 1944 workers at the West Virginia Pulp and Paper Company plant ousted Brotherhood Local 152 in favor of a CIO Paper Workers' Organizing Committee unit. See also Zieger, "The Union Comes to Covington," 69–77.

33. On contract reopenings, see Burke to Crowell Pierce, September 27, 1941, Reel 7, and to John Fleshman and others, Oct. 25, 1941, Reel 4. On CIO and AFL forays, see, e.g., file for Covington Local 152, 1941, Reel 4, and file for Vice-President Fred Morris, Sept.-Dec., 1941, Reel 1. Also Tonelli to Burke, Nov. 17, 1941, Reel 3. Burke's quoted words appear in his letter of Nov. 13, 1941, to Morris Wray and others, Reel 3.

9

Legacy

T
HE EXTENT OF THE PULP, SULPHITE, AND PAPER MILL WORKERS'
achievement in the propitious days of the 1930s remained
a matter of sharp disagreement among unionists as time
passed. For men such as Burke, Sullivan, and Stephan, the revitalization of their organization from its near-moribund condition in 1933 was nothing short of miraculous. They were proud
of the stable, multiplant agreements in the South and on the Pacific coast and remarked with wonder at the rise of unionism
among converting plant workers, rarely active before in their organization. The steady building of wage rates and narrowing of
differentials; the expansion into every geographical section; the
recruitment of black workers, urbanites, and women; the relatively
strike-free and seemingly harmonious relationships with powerful manufacturing interests — all of these phenomena proved the
validity of their cautious methods and restrained rhetoric.

Critics were not impressed. Far from finding the gradual expansion of unionism under Brotherhood auspices a contribution to
the well-being of paper workers, they viewed it as an impediment
to more effective unionism. Had there been no Pulp, Sulphite,
and Paper Mill Workers, these critics implied, organization in the
industry might well have been more thorough and dynamic. The
militancy of pulp and paper workers could have found expression

through vigorous grass-roots unionism such as that exhibited by auto and rubber workers. Had the Brotherhood not been on the scene, some CIO critics believed, pulp and paper might well have joined the other great mass production industries in the heart of the movement spearheaded by the CIO. Thus early in 1944 a CIO analysis noted with disdain that "after about 40 years of organization these two A.F. of L. unions [i.e., the Brotherhood and the Paper Makers] had organized only 70,000 workers [*sic*] out of a total of more than 500,000" workers in the industry. Organizers for CIO bodies and campaigners in pulp and paper for United Mine Workers District 50 believed that the fact that employers had the option of collaboration with the conservative AFL organizations blunted the impact of their own efforts, thus impeding the vast expansion of organization that they believed theoretically possible.[1]

The overall performance of the CIO in paper, however, did little to demonstrate the validity of this view. From the beginning of the new federation, its efforts in paper were sporadic, uncoordinated, and episodic. In New York Local 65 of the Retail, Wholesale union, a significant CIO affiliate in that city, did battle with Tonelli and company throughout the late 1930s and early 1940s. From time to time, various CIO local industrial unions representing paper workers sprang up in widely scattered locations while United Mine Workers District 50 made periodic spectacular raids and forays into pulp and paper mills. In 1938, the CIO chartered the Playthings and Novelty Workers union, which claimed some locals in urban converting plants. Two years later, local industrial unions in paper were added to the organization, and its name was changed to the Paper, Novelty, and Toy Workers. At about this time, John L. Lewis began the process that culminated in 1942 with the UMW's disaffiliation from CIO. This action in turn rendered the often vigorous District 50 sallies into pulp and paper mills a threat to the CIO as well as to the older AFL unions, notably the Pulp, Sulphite, and Paper Mill Workers. While CIO organizations occasionally gained dramatic victories, as in District 50's 1941 ouster of the AFL unions in a large mill in Berlin, New

1. Graham, *Paper Rebellion*, 19–20; Brotslaw, "Trade Unionism," 177; interview with Prof. George W. Brooks, Feb. 7, 1979; CIO Staff Memorandum, Feb. 2, 1944. The figures in the memorandum grossly understated Brotherhood membership, which in 1941 stood at more than 60,000.

Hampshire, before the war theirs was a very sporadic presence in the industry.[2]

During World War II, however, the CIO seemed about to achieve a major breakthrough. At first the new amalgamation of paper workers into the Paper, Novelty, and Toy Workers only further hindered CIO in the industry. In 1942 and 1943 disaffected locals representing paper workers threatened to bolt in favor of District 50 if the CIO did not establish a separate paper workers' organization. Faced with this "ultimatum" (the word used by Allan Haywood, the CIO director of organization) on January 1, 1944, the CIO chartered the Paper Workers Organizing Committee (PWOC), with Haywood as chairman and veteran CIO paper unionists Harry B. Sayre and Frank Grasso playing prominent roles. Freed now of encumbrance by the playthings and novelty workers' organization and filled with missionary zeal, PWOC organizers began a major effort to make the CIO an enduring presence in this large industry.[3]

PWOC's first major target was a three-mill unit of the West Virginia Pulp and Paper Company in Pennsylvania, Maryland, and Virginia which employed about 3,500 workers and was under contract to Brotherhood locals. Throughout the summer of 1944, PWOC conducted a vigorous grass-roots campaign against these Brotherhood organizations. True to form, Burke and his associates had stressed accommodation and cooperation with the company. Having endured years of antilabor harassment and delay in their efforts to secure contracts with the company, they cherished the agreements that they had reached just before World War II. With District 50 also waging an intensive campaign, PWOC forces attacked the Brotherhood's collaborationist stance and poor record in grievance handling, making particular efforts to attract strategically located black workers. In a runoff election on August 30, PWOC narrowly defeated the Brotherhood and went on to build vigorous and effective local unions in the West Virginia Pulp and Paper

2. Graham, *Paper Rebellion*, 19–21; "United Paperworkers of America," in Fink, ed., *Labor Unions*, 279–80.

3. "United Paperworkers of America," 279–80; Minutes, General Executive Board Meeting, Paper, Novelty, and Toy Workers, Nov. 7, 1943, CIO Secretary-Treasurer Papers, Box 58, Folder: Playthings, Archives of Labor History and Urban Affairs, Wayne State University (hereafter cited as ALHUA); Alex Bail, President of Playthings, Jewelry, and Novelty Workers, to Walter P. Reuther, Feb. 3, 1953, CIO Washington Office Papers, Box 26, Folder 11, ALHUA.

mills. With this victory under its belt, PWOC seemed to have lim-
itless possibilities. Secretary Harry B. Sayre declared that the new
organization was poised to "hit the jackpot" in the industry.[4]

But once again, a CIO breakthrough in the paper industry proved
elusive. Despite early rapid growth fueled by the West Virginia
victory and by successes in paper converting in the New York City
area, the United Paperworkers of America (which PWOC became
in 1946) added only slowly to its membership. It grew from just
above 18,000 after the West Virginia victory to more than 32,000
by 1947, but further expansion proved difficult. On the eve of the
CIO's merger with the AFL in 1955, the United Paperworkers num-
bered fewer than 40,000 members, a disappointing showing on
the part of the activists who in 1944 had so scornfully assessed
the Brotherhood's achievements.[5]

The two AFL organizations, and especially the Pulp, Sulphite,
and Paper Mill Workers, fared much better. The Brotherhood
reached the 100,000 mark in 1946–47, surpassed 125,000 in 1950,
and at the time of merger boasted more than 160,000 members.
In good part, the union's success resulted from adherence to policies that accompanied its rebirth in the 1930s. Certainly, as large
multiplant employers built new facilities and increasingly integrated
converting operations into their primary paper production, they
continued to view Brotherhood operations favorably, especially
if they were confronted with threats from the United Paperworkers. In addition, the 1944 defeat shocked the aging leadership and
brought about reforms in the organization's efforts to maintain
contact with its local unions. Moreover, the poor record of the
CIO's planned organizing crusade in the South, "Operation Dixie,"
dealt a severe blow to the United Paperworkers. Southern paper
production continued to mushroom, and the Brotherhood's success, as well as the sorry failure of the CIO, left the United Paperworkers with few locals south of Virginia.[6]

4. Zieger, "The Union Comes to Covington," 72–77; Harry B. Sayre to PWOC,
Aug. 31, 1944, CIO Secretary-Treasurer Papers, Box 58, Folder: Paperworkers.

5. Membership figures quoted in Graham, *Paper Rebellion*, 21; Sayre memorandum on the paper industry and the United Paperworkers of America (UPA), Jan.
30, 1953, CIO Washington Office Papers, Box 26, Folder 10; Sayre to Robert Oliver,
n.d. but c. Sept. 1953, CIO Washington Office Papers, Box 40, Folder 8.

6. Graham, *Paper Rebellion*, 21; Prof. George W. Brooks to author, Aug. 24,
1982.

But the Pulp, Sulphite, and Paper Mill Workers did not lack tribulations in the postwar decades. The legacies of the 1930s remained potent throughout the 1950s and 1960s. The United Paperworkers mounted few successful campaigns against Brotherhood organizations in the South, for example, but the Brotherhood in that region gained an unenviable reputation in the labor movement for its cozy bargaining relationships with the paper companies, for timid grievance handling, and for having apathetic and somnolent local unions.[7]

In addition, problems that had first arisen in the 1930s came home to roost in the early 1960s. A complex membership revolt, centering on the West Coast locals, challenged the union's septuagenarian leadership. The revolt stemmed from traditional western complaints of poor servicing, lack of regional representation, and belief that the international union had become increasingly dictatorial and high-handed in its dealings with the locals. In addition, resentment targeted on Joseph Tonelli, increasingly evident as John P. Burke's heir apparent. In the late 1950s, union dissidents had charged Tonelli with corrupt dealings involving certain East Coast locals. While he persuaded the union convention on two occasions that he was innocent of wrongdoing, his accusers remained popular figures in the organization. Their stature grew in 1960 when the international executive board dismissed two of Tonelli's most vocal critics from their positions as union representatives.

Adding to the increasingly tangled and seemingly sordid affairs of the organization was a controversy over the resignation of the Brotherhood's education director, George W. Brooks. Though he was a highly regarded figure in labor education activities, Brooks was an outspoken critic of some elements in the union leadership. Some members of the executive board believed that he had encouraged both the West Coast dissenters and Tonelli's critics. When Brooks resigned under pressure early in 1961, reform forces began active organization to oppose what they saw as an executive board increasingly out of touch with the membership and

---

7. Interview with Clarence Farmer, Covington, Virginia, June 4, 1980; interview with Nicholas Vrataric, Ann Arbor, Michigan, May 9, 1980; report of CIO membership by state and congressional district, Jan. 8, 1954, CIO Washington Office Papers, Box 84, Folder 1.

increasingly willing to resort to intimidation and chicanery to maintain itself while hiding behind the figure of the venerable Burke, whom they regarded as somewhat distracted. In 1964, the West Coast revolt culminated in a bitter rupture when forty-nine Pacific Northwest locals seceded to form the Association of Western Pulp and Paper Workers, thus ending the much-praised regional bargaining system that Burke and his associates had so proudly erected in the thirties.[8]

Nor did the troubles end there. In 1972, the Brotherhood merged with the United Papermakers and Paperworkers, itself an amalgamation in 1957 of the United Paperworkers and the Paper Makers. Tonelli, having succeeded Burke to the Brotherhood presidency on the latter's retirement in 1965, was elected president of the merged organization, the United Paperworkers International Union. In 1974, however, most Canadian locals, among them old loyalist bodies dating back to the 1920s, pulled away to establish a Canadian pulp and paper workers' organization. Then, in 1978, Tonelli was convicted for embezzlement of union funds and was sentenced to federal prison.[9] John P. Burke did not live to witness the disgrace of his successor. Afflicted by age and illness, in January 1965 he resigned the position of president-secretary, which he had held continuously since 1917. On April 22, 1966, he died, leaving the stewardship of the organization in Tonelli's hands.

The year 1981 marked the seventy-fifth anniversary of the founding of the International Brotherhood of Pulp, Sulphite, and Paper Mill Workers. As with so many labor unions, the Pulp, Sulphite Workers had been deeply affected by the events of the New Deal era. It had been barely clinging to life in 1933, but by 1941 the Brotherhood had become a major force in its industry, had moved with pulp and paper production into new areas, and was poised for further expansion. In the 1930s, Burke and his lieutenants seized, however diffidently at first, the opportunity to establish the organization in the mushrooming converting industry. They forged permanent bargaining relationships with International

8. Graham, *Paper Rebellion*; "Association of Western Pulp and Paper Workers," in Fink, ed., *Labor Unions*, 307–10; "Rank and File Movement for Democratic Action – International Brotherhood of Pulp, Sulphite, and Paper Mill Workers," clipping c. Jan. 1962 in Union Democracy in Action Collection, ALHUA.

9. *New York Times*, July 20, Dec. 7, Dec. 20, 1978; *John Herling's Labor Letter* 28 (Oct. 28, 1978); interview with Vrataric; interview with Brooks.

Paper, Southern Kraft, Crown-Willamette, and other major producers. Membership grew fifteenfold, providing the crucial basis
for subsequent expansion.

Critics might well question the Pulp, Sulphite, and Paper Mill
Workers' obsession with caution and restraint. Opponents certainly
scored heavily when they noted the AFL organization's obtuse policies and attitudes toward its black members and toward others
it sought to recruit. Critics also charged that the Brotherhood's
creaky hierarchical structure, combined with its distrust of local
unionists, failed to tap the energy and potential enthusiasm of
rank-and-file paper workers. Still, when allegedly more aggressive
organizations flying the magic CIO banner did eventually move
into the field, they compiled only a mediocre record.

The main thing, John Burke always believed, was to maintain
the union, to keep it alive, to defend it from rivals. Growth was
less important than survival, dramatic gains less likely than steady
improvement. The task of establishing and maintaining viable labor organizations was enormously difficult. If growth and security required governmental assistance, the prudent leader accepted
it, though he was fully aware that what the government offered
it could also shape to its own purposes. If the reluctance of workers to pay dues regularly resulted in the union's need to barter
militancy for contract security, that consequence had to be accepted. If disputatious and/or apathetic local unions required the
concentration of power in the international union apparatus, so
be it. The union must endure, even if it had to employ methods
that limited its scope and blunted its force.

Though far from charismatic, Burke could on occasion address
his enemies defiantly. After the New Deal, such opponents were
unlikely to be the employers with whom the union negotiated,
for in collective bargaining the rules were clear and the terms of
the contest increasingly well established and regulated. Rather,
the forces in the labor movement that threatened the old union
drew his scorn and invective. In 1950, workers at the Brown Brothers' mill in Berlin, New Hampshire, voted to abandon District 50
and to rejoin the Brotherhood, reversing their vote of nine years
earlier. Relishing this recapture of 2,700 members from John L.
Lewis's minions, Burke permitted himself a moment of boldness.
"We accept the challenge of every rival union," he declaimed to

delegates at the 1950 convention. "They haven't destroyed us; they are not going to destroy us."[10]

Despite Burke's death, despite the revolt of the West Coast and Canadian locals, despite the sordid and damaging behavior of Burke's own protégé, Joseph Tonelli, the union did indeed endure. The organization that Burke had guided with such caution through the heady but dangerous days of the 1930s remained at the heart of unionism in the industry, now through the United Paperworkers International. The legacy of the 1930s has been a tenacious one indeed.

10. Burke, quoted in Graham, *Paper Rebellion*, 23.

Bibliographical Essay

Works and Sources on the Pulp and Paper Industry

The most useful published work on the paper industry is David C. Smith, *History of Papermaking in America (1691–1969)* (New York: Lockwood, 1970). Smith treats the industry largely "from the perspective of the company executive" but includes a useful chapter on labor relations. The book is based extensively on company archives and publications. Valuable for statistical information and for West Coast developments is John A. Guthrie, *The Economics of Pulp and Paper* (Pullman: State College of Washington, 1950). Robert M. MacDonald, "Pulp and Paper," in Lloyd G. Reynolds and Cynthia H. Taft, eds., *The Evolution of Wage Structure* (New Haven: Yale Univ. Press, 1956), 99–166, supplies important information on union wage policies. Herbert R. Northrup, *The Negro in the Paper Industry* (Philadelphia: Wharton School, Univ. of Pennsylvania Press, 1969), and Harry J. Bettendorf, *Paperboard and Paperboard Containers: A History* (Chicago: Board Products, 1946), a surprisingly charming little book, are both helpful beyond their immediate subjects. L. Ethan Ellis, *Newsprint: Producers, Publishers, Political Pressures* (including the text of *Print Paper Pendulum: Group Pressures and the Price of Newsprint*, originally published in 1948) (New Brunswick: Rutgers Univ. Press, 1960), Wilbur F. Howell, *A History of the Corrugated Shipping Container Industry in the United States* (Camden, N.J.: Langston, 1940), and *1848–1948: A History of the Wisconsin Paper Industry* (Chicago: Howard, 1948), are useful specialized studies. My footnotes in chapter 2 contain citations for specialized reports produced under U.S. government auspices describing and analyzing pulp and paper industry economic trends, employment patterns and conditions, and related matters.

There is no overall history of the International Brotherhood of Pulp, Sulphite, and Paper Mill Workers. Irving Brotslaw, "Trade Unionism in the Pulp and Paper Industry" (Ph.D. diss., Univ. of Wisconsin, 1964), is a solid account by an economist, stressing developments after 1945. Harry Edward Graham, *The Paper Rebellion: Development and Upheaval in Pulp and Paper Unionism* (Iowa City: Univ. of Iowa, 1970), focuses on dissidence in the union after World War II, which culminated in the formation in 1964 of the Association of Western Pulp and Paper Workers. Clark

Kerr and Roger Randall, *Collective Bargaining in the Pacific Coast Pulp and Paper Industry* (Philadelphia: Wharton School, Univ. of Pennsylvania Press, 1948), is a highly laudatory account of the West Coast bargaining arrangements.

Two doctoral dissertations are useful for the predepression period. Keith Emory Voelker, "The History of the International Brotherhood of Pulp, Sulphite, and Paper Mill Workers from 1906 to 1929: A Case Study in Industrial Unionism before the Great Depression" (Ph.D. diss., Univ. of Wisconsin, 1969), is a careful study in the Commons-Perlman tradition. Stephen Rea Cernek, "Beyond the Return to Normalcy: The Decline of Organized Paper Workers, 1921–1926" (Ph.D. diss., Ball State Univ., 1978), takes a dim view of the Brotherhood's conduct in the 1920s. James A. Gross, "The Making and Shaping of Unionism in the Pulp and Paper Industry," *Labor History* 5 (Spring 1964), 183–208, is a hasty overview, touching very lightly on the 1930s. See also my articles: "The Limits of Militancy: Organizing Paper Workers, 1933–1935," *Journal of American History* 63 (Dec. 1976), 638–57; "Oldtimers and Newcomers: Change and Continuity in the Pulp, Sulphite Union in the 1930s," *Journal of Forest History* 21 (Oct. 1977), 188–201; and "The Union Comes to Covington: Virginia Paperworkers Organize, 1933–1952," *Proceedings of the American Philosophical Society* 126 (1982), 51–89. The article in the *Journal of Forest History* contains photographs of paper mills, machinery, operations performed by converting workers, and union facilities and functionaries.

These articles and the present book are based on the Papers of the International Brotherhood of Pulp, Sulphite, and Paper Mill Workers, which I used in microfilm form. These papers are housed in both the State Historical Society of Wisconsin and the Labor-Management Documentation Center, Martin P. Catherwood Library, New York State School of Industrial and Labor Relations, Cornell University. There are sixty-five reels for the period 1929–41. The Brotherhood's papers constitute a rich record of the union's activities. In addition to files relating directly to President-Secretary Burke, there are substantial files for each local union and for each officer and organizer. Particularly useful files deal with the union's relationship with companies, government bodies, and other labor organizations. The collection abounds with incoming and outgoing correspondence as well as with reports, minutes, memoranda, and other documents. Supplementing these very vivid and compendious records is the Pulp, Sulphite, and Paper Mill Workers' *Journal*, which contains photographs of local activists and organizers plus reports from local unions and international representatives. The *Journal* also includes important documents, such as Burke's major addresses, and local workers' ruminations and poetry. The *Proceedings* of the Brotherhood, as well as the various editions of its constitution, are available in micro-film form (Convention Proceedings and Constitutions of American Trade Unions, Reel 148).

Other union materials were also helpful. The American Federation of Labor collection at the State Historical Society of Wisconsin con-tains information in Series 4 (Industries File) on the converted paper industry and on government activities. The AFL-CIO records, formerly housed in the library at the AFL-CIO building in Washington and slated for inclusion in the new Meany Labor Studies Center in Silver Spring, are particularly useful in tracking down material relating to jurisdictional matters. The Archives of Labor History and Urban Affairs, Wayne State University, has significant material on the CIO organizations in paper.

I made no particular effort to examine company records, relying largely on Smith, *History of Papermaking*, on copious NRA material (see below), and on company correspondence found in the Pulp, Sulphite, and Pa-per Mill Workers' collection. I did use the Westvaco Papers, Department of Manuscripts and Archives, Cornell University Libraries, and I found them highly informative for the NRA period. My friend and assistant Barbara Hudy Bartkowiak examined files in the Marathon Paper Com-pany records at the State Historical Society of Wisconsin for me and provided useful material on the NRA period.

Government archives also yielded important information. The rec-ords of the National Recovery Administration (RG 9), housed at the National Archives in Washington, contain hearings, code histories, sta-tistical material, and correspondence indispensable in understanding the background of the paper industry and the early stages of the union's revival. The records of the National Labor Relations Board (RG 25), at the National Archives and Records Center in Suitland, Md., provide material especially illuminating on the union's activities in the South and Middle West. They are supplemented by the records of the U.S. Con-ciliation Service (RG 280), National Archives and Records Service Cen-ter, which cast light on organizing activities and employers' resistance.

Works on Labor in the 1930s

The literature on the labor movement of the 1930s is vast and ever growing. Standard histories are Irving Bernstein, *The Turbulent Years: A History of the American Worker, 1933–1941* (Boston: Houghton Mif-flin, 1969), and Walter Galenson, *The CIO Challenge to the AFL: A His-tory of American Labor, 1935–1941* (Cambridge, Mass.: Harvard Univ. Press, 1960). Art Preis, *Labor's Giant Step: Twenty Years of the CIO*, 2d ed. (New York: Pathfinder Press, 1972), is a provocative account. The es-says in Milton Derber and Edwin Young, eds., *Labor and the New Deal*

(Madison: Univ. of Wisconsin Press, 1957), are still valuable. James O. Morris, *Conflict within the AFL: A Study of Craft Versus Industrial Unionism, 1901–1938* (Ithaca: Cornell Univ. Press, 1958), remains basic. James Green, "Working Class Militancy in the Depression," *Radical America* 6 (Nov.–Dec. 1972), 1–36, and Melvyn Dubofsky, "Not So 'Turbulent Years': Another Look at the American 1930's," *Amerikastudien/American Studies* 24:1 (1979), 5–20, offer divergent judgments. See also Green, *The World of the Worker: Labor in Twentieth-Century America* (New York: Hill and Wang, 1980), ch. 5. Frances Fox Piven and Richard Cloward, *Poor People's Movements: How They Succeed, Why They Fail* (New York: Pantheon, 1977), places labor's rise in the broad perspective of a half century of reform. David Brody's reviews and essays on the 1930s, most of which appear in *Workers in Industrial America: Essays on the Twentieth Century Struggle* (New York: Oxford Univ. Press, 1980), are uncommonly illuminating.

Much of the labor history of the thirties has been written in terms of individual unions and industries. Recent research has centered on the CIO organizations that arose in the decade. Sidney Fine, *Sit-Down: The General Motors Strike of 1936–37* (Ann Arbor: Univ. of Michigan Press, 1969), is a vivid account of the auto workers' decisive strike. Peter Friedlander, *The Emergence of a UAW Local, 1936–39: A Study in Class and Culture* (Pittsburgh: Univ. of Pittsburgh Press, 1975), examines the dynamics of union genesis. Two articles by Ray Boryczka ("Seasons of Discontent: Auto Union Factionalism and the Motor Products Strike of 1935–1936," *Michigan History* 61 [Spring 1977], 3–32, and "Militancy and Factionalism in the United Autoworkers, 1937–1941," *Maryland Historian* 8 [Fall 1977], 13–25) reveal the complexity of early organization. Daniel Nelson, "Origins of the Sit-Down Era: Worker Militancy and Innovation in the Rubber Industry, 1934–38," *Labor History* 23 (Spring 1982), 198–225, deepens our understanding of a key CIO organization. Ronald Schatz, "Union Pioneers: The Founders of Local Unions at General Electric and Westinghouse, 1933–1937," *Journal of American History* 66 (Dec. 1979), 586–602, sheds light on another major CIO affiliate, employing, as does Nelson, unique oral history material. Donald Sofchalk, "The Little Steel Strike of 1937" (Ph.D. diss., Ohio State Univ., 1961), is superb on the steel workers' struggles.

AFL organizations have drawn less extensive treatment. See Galenson's individual chapters, supplemented by Philip A. Taft, *The A.F. of L.: From the Death of Gompers to the Merger* (New York: Harper, 1959), and Taft, *Organized Labor in American History* (New York: Harper, 1964). Lewis L. Lorwin, *The American Federation of Labor: History, Policies, Prospects* (Washington: Brookings, 1933), remains useful. Harry A. Millis and Royal Montgomery, *Organized Labor* (New York: McGraw-Hill, 1945), is invaluable on the mechanics of labor organizations. Monographs on key AFL unions provide additional material even when the focus is not primarily on the thirties. See especially David Brody, *The Butcher Workmen: A Study of Unionization* (Cambridge, Mass.: Harvard Univ. Press, 1964); Robert Christie, *Empire in Wood: A History of the Carpenters Union* (Ithaca: Cornell Univ. Press, 1956); Estelle James and Ralph James, *Hoffa and the Teamsters: A Study of Union Power* (Princeton: Van Nostrand, 1965); Donald Garnel, *The Rise of Teamster Power in the West* (Berkeley: Univ. of California Press, 1971); and Vernon Jensen, *Lumber and Labor* (1945; reprint ed., New York: Arno, 1971). Christopher Tomlins, "AFL Unions in the 1930s: Their Performance in Historical Perspective," *Journal of American History* 65 (March 1979), 1021–42, is particularly illuminating. Robert H. Zieger, *Madison's Battery Workers, 1934–1952: A History of Federal Labor Union 19587* (Ithaca: New York State School of Industrial and Labor Relations, 1977), deals with an oft-neglected aspect of the Federation's activities.

Other works on the unions in the 1930s that help place the experiences of the Pulp, Sulphite, and Paper Mill Workers in context include Thomas R. Brooks, *Communications Workers of America: The Story of a Union* (New York: Mason/Charter, 1977), John N. Schacht, "Toward Industrial Unionism: Bell Telephone Workers and Company Unions, 1919–1937," *Labor History* 16 (Winter 1975), 5–36, and William Haskett, "Ideological Radicals, the American Federation of Labor, and Federal Labor Policy in the Strikes of 1934" (Ph.D. diss., Univ. of California, Los Angeles, 1957).

Remote from the concerns of most pulp and paper workers but of great importance generally, especially in the CIO, were the activities of radicals. In particular, the role of the Communist party and of individuals associated with it has undergone much recent reexamination. See particularly Bert Cochran, *Labor and Communism: The Conflict That Shaped American Unions* (Princeton: Princeton Univ. Press, 1977), which is sharply critical of the Communists from a radical perspective, and Harvey A. Levenstein, *Communism, Anticommunism, and the CIO* (Westport: Greenwood, 1981), a more sympathetic study. Roger R. Keeran, *The Communist Party and the Autoworkers Unions* (Bloomington: Indiana Univ. Press, 1980), is a vigorous defense of party policies. Max Kampelman, *The Communist Party vs. the CIO* (New York: Praeger, 1957), is critical from a liberal perspective but is dated and was based on inadequate sources. Irving Howe and Lewis Coser, *The American Communist Party: A Critical History*, rev. ed. (New York: Praeger, 1962), is a more thoughtful, if equally engaged, analysis.

Autobiographies and memoirs are particularly rich and provocative on this subject. Al Richmond, *A Long View from the Left* (Boston: Houghton Mifflin, 1973); John Williamson, *Dangerous Scot* (New York: International, 1969); Wyndham Mortimer, *Organize! My Life as a Union Man,* ed. Leo Fenster (Boston: Beacon, 1971); Len De Caux, *Labor Radical: From the Wobblies to the CIO, A Personal History* (Boston: Beacon, 1970); and Steve Nelson, James R. Barrett, and Rob Ruck, *Steve Nelson: American Radical* (Pittsburgh: Univ. of Pittsburgh Press, 1981), present the stories of those in or close to the party quite forcefully. They can be supplemented with several important biographies. See Charles Larrowe, *Harry Bridges: The Rise and Fall of Radical Labor in the United States* (New York: Lawrence Hill, 1972); and Phillip Bonofsky, *Brother Bill McKie: Building the Union at Ford* (New York: International, 1953). Christopher Johnson's forthcoming biography of United Auto Workers' legal counsel and radical activist Maurice Sugar is rich in detail on the left-wing milieu in which the Detroit Communist party flourished.

Other biographies are worthy of note, although the genre is generally disappointing in its yield for labor in the 1930s. By far the best is Melvyn Dubofsky and Warren Van Tine, *John L. Lewis: A Biography* (New York: Quadrangle, 1977). Other biographies of Lewis reflect the controversies that surrounded labor's most dramatic figure. See James Wechsler, *Labor Baron: A Portrait of John L. Lewis* (New York: Morrow, 1944); and Saul Alinsky, *John L. Lewis: An Unauthorized Biography* (New York: Putnam's, 1949). The judicious reader can glean useful information from Matthew Josephson, *Sidney Hillman: Statesman of American Labor* (Garden City: Doubleday, 1952); Frank Cormier and William Eaton, *Reuther* (Englewood Cliffs, N.J.: Prentice-Hall, 1970); Thomas R. Brooks, *Clint: The Biography of a Labor Intellectual, Clinton S. Golden* (New York: Atheneum, 1978), and even Maxwell G. Raddock, *Portrait of an American Labor Leader: William L. Hutcheson: Saga of the United Brotherhood of Carpenters and Joiners, 1881–1954* (New York: American Institute of Social Science, 1955). Victor Reuther, *The Brothers Reuther and the Story of the UAW: A Memoir* (Boston: Houghton Mifflin, 1976), is an important document. See also Irving Howe and B. J. Widick, *The UAW and Walter Reuther* (New York: Random House, 1949), an insightful early study. Lorin Lee Cary, "Institutionalized Conservatism in the Early c.i.o.: Adolph Germer, a Case Study," *Labor History* 13 (Fall 1972), 475–504, is important. John Bernard, *Walter Reuther and the Rise of the Auto Workers* (Boston: Little, Brown, 1983) is an excellent account.

Many important major figures and most secondary leaders have received no adequate biographical treatment. Philip Murray, James B. Carey, Van A. Bittner, David Dubinsky, William Green, William Hutcheson, John P. Frey, Matthew Woll, and John Brophy are among those needing scholarly biographies. Oral histories of major figures and increasingly of secondary and rank-and-file activists are available at Columbia, Pennsylvania State, and Wayne State universities and elsewhere. Also helping to fill the biographical void is Murray Kempton, *Part of Our Time: Some Monuments and Ruins of the Thirties* (1955; reprint ed., New York: Delta, 1967), a series of incisive sketches, some of laborites.

The role of black workers has been receiving increasing, if belated, attention. William H. Harris, *The Harder We Run: Black Workers since the Civil War* (New York: Oxford Univ. Press, 1982), is a good overview. Harris, *Keeping the Faith: A Philip Randolph, Milton P. Webster, and the Brotherhood of Sleeping Car Porters* (Urbana: Univ. of Illinois Press, 1977), and Jervis Anderson, *A. Philip Randolph: A Biographical Portrait* (New York: Harvest, 1973), chronicle the tribulations of black leaders in the AFL, as does Marc Karson and Ronald Radosh, "The American Federation of Labor and the Negro Worker," in Julius Jacobson, ed., *The Negro and the American Labor Movement* (New York: Anchor, 1968), 155–87. Sumner N. Rosen, "The CIO Era, 1935–55," in Jacobson, *The Negro and the American Labor Movement*, and Philip Foner, *Organized Labor and the Black Worker, 1619-1973* (New York: Praeger, 1974), provide basic material on the new unionism and black workers. Horace R. Cayton and George S. Mitchell, *Black Workers and the New Unions* (Chapel Hill: Univ. of North Carolina Press, 1939), is still fresh and informative, while August Meier and Elliott Rudwick, *Black Detroit and the Rise of the UAW* (New York: Oxford Univ. Press, 1979), contains extensive material on a key industrial center. James Olson, "Organized Black Leadership and Industrial Unionism: The Racial Response, 1936–45," *Labor History* 10 (Summer 1969), 475–86, like the Meier and Rudwick volume, focuses on the emerging black-CIO coalition. Harvard Sitkoff, *A New Deal for Blacks: The Emergence of Civil Rights as a National Issue: The Depression Decade* (New York: Oxford Univ. Press, 1978), and Raymond Wolters, *Negroes and the Great Depression: The Problem of Economic Recovery* (Westport: Greenwood, 1970), provide both broad context and good treatment of labor issues. Herbert R. Northrup's series on black workers in various industries (Philadelphia: Wharton School, Univ. of Pennsylvania Press, 1968–74) provides much useful material. Stanley Greenberg, *Race and State in Capitalist Development* (New Haven: Yale Univ. Press, 1980), contains excellent accounts of black workers in Alabama at various points in the period after the Civil War and pays particular attention to black workers and the labor movement.

The rise of the labor movement in the 1930s was inextricably bound up with political and governmental matters. Arthur M. Schlesinger, Jr., *The Age of Roosevelt*, 3 vols. (Boston: Houghton Mifflin, 1957–60), remains the best general introduction and contains vivid material on the New Deal and labor. Frank Freidel, *Franklin D. Roosevelt*, vol. IV: *Launching the New Deal* (Boston: Little, Brown, 1973), James MacGregor Burns,

Roosevelt: The Lion and the Fox (New York: Harvest, 1956), and William E. Leuchtenburg, *Franklin D. Roosevelt and the New Deal, 1932–1940* (New York: Harper, 1963), must be consulted. Biographies of the two congressional figures most important for organized labor in the 1930s are excellent. See Joseph J. Huthmacher, *Senator Robert F. Wagner and the Rise of Urban Liberalism* (New York: Atheneum, 1968), and Patrick J. Maney, *"Young Bob" La Follette: A Biography of Robert M. LaFollette, Jr., 1895–1953* (Columbia: Univ. of Missouri Press, 1978). Jerold S. Auerbach, *Labor and Liberty: The LaFollette Committee and the New Deal* (Indianapolis: Bobbs-Merrill, 1966), establishes the context of fear and violence in which much organizing activity took place. George Martin, *Madam Secretary: A Biography of America's First Woman Cabinet Member* (Boston: Houghton Mifflin, 1976), faithfully reflects the Roosevelt administration's perspective.

The legal-bureaucratic context for labor relations that emerged in the New Deal is well chronicled. Irving Bernstein, *The New Deal Collective Bargaining Policy* (Berkeley: Univ. of California Press, 1950), remains standard. James A. Gross, *The Making of the National Labor Relations Board, 1933–1937*, vol. I (Albany: State Univ. of New York Press, 1974), and *The Reshaping of the National Labor Relations Board: National Labor Policy in Transition, 1937–1947* (Albany: State Univ. of New York Press, 1981), are heavily documented. The 1981 volume is particularly useful for its coverage of the AFL-supported attack on the board in the late 1930s. Cletus Daniel, *The ACLU and the Wagner Act* (Ithaca: New York State School of Industrial and Labor Relations, 1981), examines radical criticism of the Labor Relations Act. Richard Cortner, *The Wagner Act Cases* (Knoxville: Univ. of Tennessee Press, 1964), analyzes constitutional issues. Ellis W. Hawley, *The New Deal and the Problem of Monopoly: A Study in Economic Ambivalence* (Princeton: Princeton Univ. Press, 1966), is the standard study of the NRA, while Sidney Fine, *The Automobile under the Blue Eagle: Labor, Management, and the Automobile Manufacturing Code* (Ann Arbor: Univ. of Michigan Press, 1963), is a richly detailed analysis of the NRA's failures in the labor field.

The view that the New Deal reforms were clearly and substantially prolabor in impact, if not always in explicit intent, is well stated in works by Schlesinger, Leuchtenburg, Bernstein, and Burns, cited above. See also Richard Hofstadter, *The Age of Reform: From Bryan to F.D.R.* (New York: Knopf, 1955), and Michael Harrington, *Socialism* (New York: Bantam, 1972), ch. 11. Barton J. Bernstein, "The New Deal: The Conservative Achievements of Liberal Reform," in Bernstein, ed., *Towards a New Past: Dissenting Essays in American History* (New York: Pantheon, 1968), 263–88, and Ronald Radosh, "The Corporate Ideology of American Labor Leaders from Gompers to Hillman," *Studies on the Left* 6 (Nov.–Dec.

1966), 66–88, stress the limits of New Deal measures. For more recent, vigorously stated dissent, see Mike Davis, "The Barren Marriage of American Labour and the Democratic Party," *New Left Review* 124 (Nov.–Dec. 1980), 43–84, and David Montgomery, "American Workers and the New Deal Formula," in Montgomery, *Workers' Control in America: Studies in the History of Work, Technology, and Labor Struggles* (Cambridge: Cambridge Univ. Press, 1980), 153–80. See Theda Skocpol, "Political Response to Capitalist Crisis: Neo-Marxist Theories of the State and the Case of the New Deal," *Politics and Society* 10 (1980), 155–201, for a broad discussion of these themes.

Index

Akron, Ohio, 9, 76, 78, 81, 119
Akron Central Labor Union, 73
Alabama, 112; Pulp, Sulphite Workers' activity in, 68, 142, 143
Allentown, Pa., 119
Amalgamated Association of Iron, Steel, and Tin Workers, 6, 12
Amalgamated Meat Cutters and Butcher Workmen of North America, 14
American Federation of Labor (AFL), 4, 6, 45, 67, 95, 130, 131, 134, 136, 138, 141, 150, 156, 158, 163, 181, 182, 187, 212, 213, 215; and black workers, 114; John Burke criticisms of, 61, 191–92; CIO challenge to, 127, 177, 181; craft union control of, 172, 182; and industrial unionism, 11, 61, 65, 180; and jurisdictional problems, 168–80, 191; and Labor Advisory Board, 84–85; attack on NLRB, 170, 192; and Pulp, Sulphite Workers' union, 10, 11, 12, 46–47, 48, 68, 72–73, 139, 169–70, 186, 217; activities in South, 143–44
American Federation of Labor Executive Council, 169, 173, 182, 183, 192; craft unions dominate, 170; jurisdictional awards by, 174–75; seeks amendments to NLRA, 96
American Plan, 49
American Pulp and Paper Association, 84
Anderson, Rasmus (Pulp, Sulphite Workers' organizer), 73
Apparel industry, 14
Association of Western Pulp and Paper Workers, formation of, 219, 221

Baltimore, 163
Barkin, Solomon, 187
Barnes, Frank C., Jr. (Pulp, Sulphite Workers' organizer), 163
Barry, Frank, 149
Bastrop, La., 147
Beach, Harry (Pulp, Sulphite Workers' organizer), 160
Beck, Dave (Pulp, Sulphite Workers' organizer), 125
Berlin, N.H., 27, 184, 210, 215–16, 220
Berry, George, 176, 204
Black workers, pulp and paper industry: Burke's attitudes toward, 115–16; CIO recruitment of, 147; discrimination against, 59, 113; importance of in pulp and paper industry, 34, 112–116; Pulp, Sulphite Workers' recruitment of, 114–16, 151, 195, 214; wages and working conditions of, 113, 200, 201
Bloomfield, N.J., 160
Blue Eagle, labor support for, 12, 85, 88
Bogalusa, La., 209
Bogota, N.J., 161
Bohn, William, 59
Boston, 152, 159
Bridges, Harry, 8
British Columbia, 140
Brodzinski, Valeria (Pulp, Sulphite Workers' organizer), 109, 119
Bronx: bag workers' strike, 1936, 99, 156–57
Brooklyn Standard Boys' and Girls' Protective Association, 74
Brooks, George C. (Pulp, Sulphite Workers' international auditor):

Brooks, George C. (*cont.*)
on impact of NRA, 68, 72, 82–83;
organizing duties of, 104, 117, 124,
159; on paper workers, 76, 82, 157;
on black workers, 111, 114–15; on
women workers, 107, 110
Brooks, George W. (Pulp, Sulphite
Workers' director of education),
218
Brookwood Labor College, 109
Bujold, Leo (local unionist), 76
Bureau of Labor Statistics: report
on living standards, 37, 70; report
on converted paper industry, 35,
39
Burke, Bessie Leon, 53, 101, 120
Burke, John P. (Pulp, Sulphite Workers' president-secretary), 8, 74, 91,
92, 113, 116, 144, 146, 164, 202, 218;
on AFL, 61, 96, 169–70, 175, 181–82,
191–92; biographical sketch of, 48–
54, 125–26, 219–20; on black
workers, 114–16, 151; on CIO, 60,
180–90; on converted paper industry, 42, 78, 79, 152–54, 158, 160,
161, 163, 165, 185; death of, 219; on
dues, union shop, 12, 57, 128–32,
147–48, 177, 178, 208–9; on foreign
policy, 203; on history and
achievements of Pulp, Sulphite
Workers, 3, 12, 52, 96, 117, 163, 195,
213, 214, 220–21; on jurisdictional
concerns, 168–80, 192, 193; de-
fends NLRB, 96, 170, 192; and NRA,
26, 66–68, 70, 73, 77, 81–84, 85–87,
90–91, 132; and New York City locals, 106, 156, 206; and organizers,
organizing strategies, 41, 45, 60–
64, 77, 98, 99, 101, 104–5, 118–19,
121, 123, 124–25, 127, 145, 163–64,
179; and Pacific Northwest bargaining, 138–41; resignation of,
219; social and political philosophy of, 58–61, 64, 104, 203–5; and
southern paper industry, 116, 146–
52; and strikes, 41, 50, 51, 99–101;
and Joseph Tonelli, 207n; on
unions, 5, 54–55, 57–58, 75, 86–87,

Burke, John P. (*cont.*)
122–23, 220; on wage policy, 37,
38, 151, 197–202; on women workers, 108–9, 162; views of paper
workers, 43, 67, 76–77, 93, 94, 111,
122, 155, 157, 158–59
Burnell, William H. (Pulp, Sulphite
Workers' vice-president), 117
Burns, Matthew, 146; on John P.
Burke, 54; on checkoff, 210; at
NRA code hearings, 132; on labor
unrest, 43; relations with paper
companies, 145, 147
Business unionism, 8

California: Pulp, Sulphite Workers'
activities in, 139–40, 152, 162–63,
177, 178–79
Camden, Ark., 89–90, 143, 147, 151
Canada, 171; newsprint production
in, 16; Pulp, Sulphite Workers'
membership in, 51–52, 133; Pulp,
Sulphite Workers' locals secede,
1974, 219, 221
Canadian-American Treaty (1913), 48
Canadian International Paper Com-
pany, 21, 145
Carey, Jeremiah, 46, 48
Carthage, N.Y., 9
Champion Fibre and Paper Com-
pany, 112, 142–43, 150
Checkoff: importance of in union
financing, 128, 178, 209–11
Chemical industry, 14
Chicago: Pulp, Sulphite Workers' ac-
tivity in, 119, 152, 158, 159, 160,
162, 166
Christianity: John P. Burke's views
on, 54
Cianciulli, Bernard (Pulp, Sulphite
Workers' organizer), 186, 207
Cincinnati, 129
Civil liberties: John P. Burke on,
203–4
Clayton Act, 63
Cleveland, Ohio: Pulp, Sulphite
Workers' activity in, 73, 76, 78,
119, 160

Coeur d'Alene, Idaho, 41

Collective bargaining, 77, 147–49, 195; John Burke critique of, 57–58, 62–64; contract, importance of, 57–58

Commerce, United States Department of: paper industry study, 17

Committee to Defend America by Aiding the Allies, 203

Committee on Industrial Organization. *See* Congress of Industrial Organizations

Commons, John R., 49

Communications Workers of America, 6

Communism: John P. Burke on, 59–60; role in CIO, 185, 190; role in Pulp, Sulphite Workers' union, 181, 211

Company unions, 49, 68, 75, 79, 85

Congress of Industrial Organizations (CIO), 4, 98, 158, 163, 192, 215, 217; and split with AFL, 127, 181; and black workers, 7; John P. Burke view of, 60, 180–82, 183, 190; character and activities of, 7, 9, 14, 61, 150, 182, 220; rivalry with Pulp, Sulphite Workers' union, 11, 13, 119, 142, 144, 146–47, 150–51, 152, 154, 159, 166–67, 168, 176, 180–90, 193, 196, 202, 206–8, 210, 212, 213, 214–17; support for in Pulp, Sulphite Workers' locals, 106, 107, 115, 139, 140, 165, 181

Container Corporation of America, 28, 154

Converted paper industry: character of, 15, 27–29, 35–36, 38, 39, 43–44, 77, 153, 154; employers' attitudes toward unions, 78–79, 91–92, 160; employment and labor force patterns, 28, 29, 36–37, 153, 155–56, 166; importance to Pulp, Sulphite Workers' union, 153, 154, 158; Pulp, Sulphite Workers' membership in, 42, 78, 82, 106, 152–66, 184–86, 205–7; wages and working condi-

Converted paper industry (*cont.*) tions in, 38–40, 70; *see also* Pulp and Paper Industry

Cook, George (Pulp, Sulphite Workers' organizer), 161

Covington, Ky., 179

Covington, Va., 80, 89, 144, 188

Cowdrill, Robert, 88–89

Craft unions: Burke critique of, 192; control of AFL by, 182; Pulp, Sulphite Workers' disputes with, 119, 137, 140–41, 146, 150, 168–80, 181, 202, 212, 213; role in strikes of 1920s, 50–51; support for Pulp, Sulphite Workers' organizing, 72

Crown-Willamette Paper Company, 137, 173, 220

Crown-Zellerbach Paper Company, 90–91, 135, 137, 195

Cullen, R.J., 145

De Leon, Daniel, 169

Debs, Eugene V., 58–59, 64, 75, 204

Delair, N.J., 125

Democratic party: John Burke's opinion of, 58, 104, 204–5

Doody, Bart (Pulp, Sulphite Workers' treasurer), 104

Dubinsky, David, 61

Dues, 110, 145; John Burke's view of, 57, 128–31; importance to workers, 4, 164, 177, 178; increase rejected, 208–9; Pacific Northwest locals seek reduction of, 139; Pulp, Sulphite Workers' policy regarding, 177; *see also* union shop

Dunn, Robert, 87

Economic policy: John Burke on, 203

Employers: John Burke critique of, 64; role in union growth, 11; *see also* converted paper industry and pulp and paper industry

Erie, Pa., 119–20

Escanaba, Mich., 125

Everest, D.C., 21, 49

Everett, Wash., 137, 140

Fair Labor Standards Act, 202
Fall River, Mass., 82, 159
Federal government: John Burke critique of, 58, 203–4; role of in 1930s, 14
Federal Labor Union 18239, 158, 159
Federal labor unions (AFL), 6, 177
Fitzgerald, James F. (Pulp, Sulphite Workers' president-secretary), 46
Flint, Mich., 10, 14, 187
Florida: Pulp, Sulphite Workers' activity in, 112, 142
Foreign policy: John Burke on, 203
Foremen: place in bargaining units, 133
Fort Edward, N.Y., 53, 101, 104, 106, 120, 133
Fortune magazine: series on pulp and paper industry, 1937, 18, 23, 28
Fourdrinier (papermaking machine), 20–21
Fox River Valley, Wis., 72, 134
Franklin, N.H., 52
Friend, J.H., 145, 146, 148–50, 172

Gastonia, N.C., 8
George and Sherrard Paper Company, 89–90
Georgetown, S.C., 143, 146, 147
Georgia, 142
Germer, Adolph, 187
Gordon, David (local unionist), 74, 106
Gordon, Hyman (Pulp, Sulphite Workers' Local 107 president), 74, 106, 156, 157, 165, 181, 206
Gordon, Lou, 106
Grasso, Frank, 216
Graustein, Jacob, 77
Great Northern Paper Company, 16, 21, 51–52, 133, 195
Green, William, 8, 170, 191
Green Bay, Wis., 56, 76, 128, 134
Grievances: Pulp, Sulphite Workers' contract provisions regarding, 196
Gulf States Paper Company, 79

Hartford City, Ind., 92, 119
Haywood, Allan S., 216
Hitler, Adolph, 58
Heron, Alexander, 138
Hoberg Paper Company, 128
Hoboken, N.J., 161
Hoff, A.B. (Pulp, Sulphite Workers' organizer), 115, 121
Hoffman, Walter E. (local unionist), 91
Holyoke, Mass., 41, 46, 108
Hoover, Herbert, 136
Hoquiam, Wash., 136, 140
Hutcheson, William, 174, 191
Hudson Falls, N.Y., 26

Ideal Corrugated Box Company, 91–92
Illinois, 119, 153
Indiana, 91
Indianapolis Regional Labor Board (NLRB-I), 88–89
Industrial unionism, 61, 137, 170, 177; *see also* jurisdictional concerns
Industrial Workers of the World, 57, 169
International Association of Machinists (AFL), 50–51, 150, 170, 172, 179, 180, 191
International Brotherhood of Boilermakers . . . (AFL), 180
International Brotherhood of Bookbinders (AFL), 175, 176, 177, 179, 181, 185
International Brotherhood of Electrical Workers (AFL), 150, 170, 179, 180, 191
International Brotherhood of Firemen and Oilers (AFL), 50, 179, 180
International Brotherhood of Paper Makers (AFL), 41, 43, 84, 191, 197, 198, 210, 219; history of, 46–47; relations with employers, 40, 116, 134, 138, 142, 145; relations with Pulp, Sulphite Workers', 25, 46, 50, 73, 180
International Brotherhood of Paper Makers, Pulp, Sulphite, and Paper Mill Workers (AFL), 46

International Brotherhood of Pulp, Sulphite, and Paper Mill Workers (AFL), 3, 14, 15, 40–41, 48, 78, 79, 92, 95, 98–99, 101, 104, 110–12, 120, 126, 131–34, 153–59, 186, 194–213, 218–19; and AFL, 176, 182–83; and black workers, 112–16; and Canadian locals, 9, 11, 120; and CIO, 13, 98, 180–90, 215; critics of, 105–6, 212, 214–15, 220; demographic composition of membership, 9, 75–76, 110–16, 194, 195, 211–21; experience in 1930s, 10–11, 65, 77, 95–96, 127–32, 182, 213, 214–15, 216; financial circumstances, 120, 178, 208–9; history (pre-1933), 9, 11, 12, 45–53, 84, 219; jurisdiction, 25, 44, 47, 152, 154, 164, 168–93; leadership of, 4, 10, 11, 52–53, 60–65, 194; local unions: no. 25 (Rumford, Maine), 76; no. 36 (Luke, Md.), 190; no. 97 (Cloquet, Minn.), 42; no. 107 (New York City), 42, 59–60, 74, 106, 155, 156, 165, 181, 206, 207; no. 145 (Akron), 76, 81; no. 150 (Rittman, Ohio), 100; no. 159 (NIRA local, Lockland, Ohio), 69, 71, 92; no. 164 (Parkersburg, W. Va.), 91–92; no. 187 (Toledo), 109–110; no. 190 (New Deal local, Greater Boston), 71; no. 203 (Holyoke, Mass.), 108; no. 229 (Mobile), 146; no. 234 (Bronx), 165, 206; no. 240 (Connellsville, Pa.), 129; no. 270 (Nashua, N.H.), 108; no. 286 (Philadelphia), 160, 164, 166, 185, 205, 210; no. 299 (New York City Paper Box Makers), 42, 96–97, 159, 164, 165, 166, 184–85, 195, 206, 207, 211; no. 323 (Shelbyville, Ind.), 188; no. 337 (Mobile), 112, 146; no. 337-A (Mobile), 112; no. 362 (Bogalusa, La.), 209; no. 381 (Greater New York Folding Box Workers), 165, 207; no. 395-A (Fernandina, Fla.), 195; no. 413 (New York City), 184, 207; membership,

International Brotherhood of Pulp, Sulphite, and Paper Mill Workers (*cont.*) 9, 12, 13, 48, 51, 69, 93, 95–96, 194, 207, 215, 217, 220; NRA, experiences under, 66–71, 81–83, 87, 90–93; organizers' activities, 72–77, 116–125; in Pacific Northwest, 135–42, 218–19; in South, 112–13, 116, 142–52; structure and governance of, 10, 60, 99–101, 101–5, 120–21, 131, 136, 205–6, 207–11

International Brotherhood of Teamsters . . . , 6, 95, 111, 123, 125, 156, 175, 176, 178, 179, 181, 191, 192

International Falls, Minn., 26, 134

International Hod Carriers . . . Union of America, 179

International Ladies Garment Workers Union, 67

International Longshoremen's Association, 137, 180

International Paper Company, 16, 21, 24, 28, 44, 63, 77, 112, 144, 151, 172; relations with Pulp, Sulphite Workers, pre-1933, 41, 50, 51, 55–56, 142–43; relations with Pulp, Sulphite Workers after 1933, 61, 133, 145–48, 149, 190, 195, 198–99; *see also* Canadian International Paper Company and Southern Kraft

International Printing Pressmen and Assistants' Union, 163, 175, 176, 177, 179, 180, 181, 185, 193

International Union of Operating Engineers, 180

International Woodworkers of America, 140

Interstate Commerce Commission, 19

Jersey City, N.J., 160, 161, 184

Jones, Mary ("Mother"), 75

Jurisdictional concerns, 4, 47, 123, 164, 168–80, 193

Kaukauna, Wis., 68

Killen, James (Pulp, Sulphite Workers' organizer), 125

Kimberly-Clark Company, 28, 34; company unionism of, 49, 68, 75, 85
Kipnis, Samuel, 186

La Follette, Robert M., 58, 104, 204
Labor Advisory Board (NRA), 84–85, 88
Labor history: John Burke on, 3–4; interpretations of, 6–7
Labor movement, 3, 4, 6, 7–8, 84–85, 204–5
Labor party: John Burke against, 204
Labor policy, federal, 4, 6, 66–94, passim
Lend lease, 203
Leon, Raymond (Pulp, Sulphite Workers' organizer), 101, 159–60
Lewis, John L., 8, 58, 61, 63, 181, 182, 190, 191, 220; and CIO, 13, 67, 142, 186, 215
Lewis, Kathryn, 191
Livermore Falls, Maine, 133
Lockland, Ohio, 69, 92
Logging industry, 14, 137
Longview, Wash., 9, 68, 77, 135, 136, 140
Los Angeles, 162–63, 166
Louisiana: Pulp, Sulphite Workers' activities in, 68–69, 91, 142, 143, 144, 209
Louy, Lambert (Pulp, Sulphite Workers' organizer), 160, 188
Ludlow Massacre, 41
Luke, Md., 189
Luke, William, 22

Machine tenders, 45–46
Madison, Maine, 9, 133
Maine: Pulp, Sulphite Workers' activity in, 16, 71, 82, 171
Malin, James (Pulp, Sulphite Workers' president-secretary and organizer), 48, 53; in southern organizing activities, 115, 146, 150
Mangan, Edward (Pulp, Sulphite Workers' organizer), 73–74, 76, 92, 119–20, 156, 163

Marathon Mills, 21, 49, 68, 80, 134
Marrero, La., 74
Marsh, E.P., 174
Maryland, 216
Massachusetts, 111, 153
Meany, George, 191
Menasha, Wis., 68, 109
Miami Valley, Ohio, 119
Michigan, 134
Militancy, 4; John Burke's view of, 63; factor in labor growth, 7, 11; NIRA stimulates, 67–71; see also Converted paper industry; International Brotherhood of Pulp, Sulphite, and Paper Mill Workers; and Pulp and paper industry
Miller, Mary (local unionist), 109–10
Millinocket, Maine, 133
Milwaukee, 78, 154, 155, 160, 162, 177, 195
Mining industry, 14
Minneapolis, 14, 177
Minnesota: Pulp, Sulphite Workers' activity in, 11, 134, 165
Mississippi, 112, 149
Mobile, 143, 146
Moncrief, Sam (local unionist), 42
Moon, Nelson (local unionist), 129
Morris, Ill., 162
Mosinee, Wis., 97
Moss Point, Miss., 146
Muir, Abe, 174

Nashua Gummed Paper Company, 108
National Container Corporation, 154, 186
National Industrial Recovery Act, 135; John Burke supports, 65; passed, 12; Pulp, Sulphite Workers' experiences under, 11, 12, 13, 45, 66–68, 71, 72; Supreme Court and, 93
National Labor Board, 87
National Labor Relations Act: impact of, 11, 13, 203; passed (1935), 93–94; Pulp, Sulphite Workers' reaction to, 95

National Labor Relations Board (1934–35), 87

National Labor Relations Board (1935–), 188; AFL seeks changes in 170, 192; Pulp, Sulphite Workers' experiences with, 96–99, 127, 134, 170

National Paper Box Manufacturers Association, 35–36

National Recovery Administration (NRA), 132, 143, 160, 162, 202; codes and hearings, 28–29, 43, 66, 70, 74, 77, 81, 82, 83, 86–87, 109; and labor, 12, 42, 71, 77, 79, 81, 82, 83, 87, 88–93, 96, 136, 142, 144, 154, 155, 157, 169, 203; termination of, 95

Nazis, 203

Neenah, Wis., 134

Nekoosa-Port Edwards, Wis., 49, 68, 97–98

Nestor, Agnes, 109, 162

New Bedford, Mass., 159

New Deal, 6, 9, 67, 212, 220; and labor, 8, 42, 43, 45, 65, 70–71, 203–4, 219

New England: Pulp, Sulphite Workers' activities in, 11, 68, 72, 90, 93, 106, 108, 119, 133, 154, 155, 201

New Jersey: Pulp, Sulphite Workers' activity in, 111, 124, 156, 161, 164, 165, 184, 213

New Leader, 54

New Orleans Regional Labor Board (NLRB-1), 89–90

New York City: Pulp, Sulphite Workers' activity in, 74, 97, 105–6, 108, 110, 111–12, 152, 154, 155, 156, 158, 159–60, 163, 164, 166, 177, 184–86, 190–91, 195, 205–6, 217

New York State, 16, 153; Pulp, Sulphite Workers' activity in, 11, 72, 90, 93, 128, 133

Newark, N.J., 160

Newfoundland, 120

Niagara Falls, N.Y., 133

North Carolina, 68

Oakland, Calif., 163

Ocean Falls, B.C., 59

Office of Production Management, 202

Ohio: Pulp, Sulphite Workers' activity in, 68, 91, 99, 106, 119, 153, 154, 165

Ohio Boxboard Company, 100

Oil industry, 14

One Big Union, 169

Oregon, 135, 139

Oregon State Federation of Labor, 137

Organizers: activities and outlook of, 4, 5, 82, 95, 101–4, 108–9, 115, 116–25, 126

Organizing: Pulp, Sulphite Workers' policy and strategy, 69, 71–75, 87, 179

Osborne, Ben T., 137

Out of the Night, 59

Overpass, Battle of the, 8

Oxford Paper Company, 134

Pacific Coast Pulp and Paper Manufacturers' Association, 138

Pacific Northwest: pulp and paper industry growth in, 18, 20, 48, 49; Pulp, Sulphite Workers' activities in, 9, 68, 72, 90–91, 93, 119, 127, 128, 134, 135–42, 151, 166, 172–74, 200, 214; Pulp, Sulphite Workers' locals dissent, 105, 139–41, 181, 219, 221

Pacific Northwest Conference of Pulp and Paper Employees, 137

Palmer, N.Y., 133

Panama City, Fla., 9, 143, 146, 148

Paper, Novelty, and Toy Workers (CIO), 190, 215, 216

Paper Workers Organizing Committee (CIO), 216–17

Parkersburg, W. Va., 91

Pasadena, Tex., 143, 150

Pawtucket, R.I., 159

Pearl Harbor, 190

Peck, Gustav, 84

Pennsylvania: Pulp, Sulphite Work-

Pennsylvania (*cont.*)
ers' activities in, 119, 153, 165, 195, 216
Per capita tax. *See* dues
Philadelphia: Pulp, Sulphite Workers' activity in, 110, 111, 152, 158, 159, 164, 165, 176–77, 185, 188, 205
Phillips, Paul, 146
Piedmont, W. Va., 22, 189
Pittsburgh: Pulp, Sulphite Workers' activity in, 108, 119, 152, 160, 177, 195
Poe, Charles, 172
Port Angeles, Wash., 136, 139
Providence, R.I., 159
Public Resolution No. 44, 87; *see also* NLRB, 1934–35
Pulp and paper industry: character and development of, 9, 10, 15–24, 28–29, 34, 40, 43–52, 83, 142–43, 179, 197–99; labor policies of employers, 13, 34–35, 40–41, 42, 44, 77, 80, 83–84, 88, 91–92, 98, 131–34, 137–39, 188–89; workers, 4, 11, 18, 23, 24–27, 29, 40, 43, 44, 46–47, 51, 52, 55–56, 61, 66–72, 81, 106–116, 121–22, 128–32, 142–43, 150–52, 178, 197–202, 215

Radicalism: a force in 1930s, 8
Rand School, 58, 59
Randolph, A. Philip, 151
Rayonier Corporation, 112, 113, 135
Recession of 1937–39, 148, 197–99
Republic Steel Massacre, 162
Republican party, 104, 204
Retail, Wholesale, and Department Store Employees, Local 65, 184–85, 215
Reuther, Walter, 8
Richards, Raymond (Pulp, Sulphite Workers' vice-president), 97–98, 104, 119, 125, 128
Rittman, Ohio, 73, 78, 100, 119
River Rouge, Mich., 8
Roanoke Rapids, N.C., 114
Robert, Nicholas Louis, 20
Robert Gair Company, 28, 195, 199

Roosevelt, Franklin D., 58, 71, 203, 204
Rothschild, Wis., 134
Rubber workers, 73

S.D. Warren Company, 34
St. Regis Paper Company, 51, 134, 195
Salem, Ore., 137
San Francisco, 14
Savannah, Ga., 114–15, 123, 143
Sawmill and Timber Workers Union, 173–74
Sawmilling industry, 14, 137
Sayre, Harry B., 216–17
Schneider, Cong. George, 104
Scott Paper Company, 28, 34, 49
Scranton Declaration, 170
Sears, Peter (local unionist), 74
Seaside, Ore., 173–74
Segal, Harry (local unionist), 210–11
Shelbyville, Ind., 99, 119, 188
Shelton, Wash., 136
Sherman, John (Pulp, Sulphite Workers' vice-president), 140; activities as organizer, 117, 118, 125, 163; elected vice-president of Pulp, Sulphite Workers, 139; member Washington state legislature, 101, 104
Smith, Paul, 73
Smith, Wilford (Pulp, Sulphite Workers' organizer), 115
Smole, John (local unionist), 76
Socialism, socialists, 53, 58, 59, 204
Socialist Trades and Labor Alliance, 169
South: pulp and paper industry expansion into, 18, 20, 21, 28, 48–49, 112, 142–43; Pulp, Sulphite Workers' activities in, 68, 72, 90–91, 112–14, 127, 128, 134, 142–52, 172, 195, 214, 218; wage rates in, 200–1
South Chicago, 8
Southern Kraft Paper Company, 112, 151, 172; CIO and, 98, 113, 183; development of, 21, 142–43, 146; rela-

Southern Kraft Paper Co. (*cont.*)
    tions with Pulp, Sulphite Work-
    ers, 116, 144, 147, 148–50, 195, 198–
    99, 200, 213, 220
Southern Kraft Workers Organizing
    Committee (CIO), 98, 113, 183
Southgate, Calif., 163
Spanish River Paper Company, 136
Steel Workers Organizing Commit-
    tee, 182
Stephan, Jacob (Pulp, Sulphite
    Workers' organizer), 61, 68, 72, 81,
    111, 125, 157, 158, 214
Stephens, S.A. (Pulp, Sulphite
    Workers' vice-president), 99–100,
    104, 117–18, 119, 164
Stevens Point, Wis., 61
Strikes: Bronx bag workers (1936),
    99, 156–57; Burke on, 92, 99–101,
    141; in 1920s, 41, 49–51, 60–61, 63,
    101, 133, 171; Rittman, Ohio (1936),
    100
Sullivan, Herbert W. (Pulp, Sulphite
    Workers' vice-president), 40, 100–1,
    136, 157, 164, 212, 214; duties as or-
    ganizer, 82, 104, 117, 125; at NRA
    code hearings, 66, 83, 85–87, 132
Sund, R.J., 80
Supreme Court, 93, 96
Switchmen's Union of North Amer-
    ica, 73

Tate, Wilmer, 73
Terre Haute, Ind., 88–89
Terre Haute Paper Company, 88–89
Textile industry, 14
Thomas, Norman, 104, 204
Thorold, Ontario, 186
Tierney, Francis, 120
Tobin, Daniel, 176, 204
Toledo, Ohio, 75–76, 78, 106, 109,
    155, 160, 177, 195
Tonelli, Joseph (Pulp, Sulphite
    Workers' organizer, president-
    secretary), activities as organizer,
    165, 184–86, 206–7; Burke's view
    of, 207; becomes union president,
    219; controversies surrounding,

Tonelli, Joseph (*cont.*)
    218–19; conviction and imprison-
    ment, 219, 221
Toronto, 204, 207
Towsen, James, 189
Trautmann, Walter (Pulp, Sulphite
    Workers' organizer), 119, 123, 125,
    162
Tuscaloosa, Ala., 79, 144
Tyronne, Pa., 189

Uniform Labor Agreement, 138–39,
    141, 166, 194, 195; Joint Relations
    Board, 138
Union Bag and Paper Company, 28,
    112, 115, 124, 142, 143
Union shop, 56–57, 178; importance
    for Pulp, Sulphite Workers, 128,
    130–31, 147, 210; role in Southern
    organizing, 150–52; workers' atti-
    tudes toward, 129
United Automobile Workers (CIO),
    6, 10, 182, 210
United Brotherhood of Carpenters
    and Joiners, 6, 51, 95; rival of
    Pulp, Sulphite Workers', 170–75,
    179, 181, 191, 192
United Mine Workers, 6, 9, 10, 67,
    187, 210, 215; rival of Pulp, Sul-
    phite Workers, 170, 191–92
United Mine Workers District 50:
    rival of Pulp, Sulphite Workers,
    13, 98, 183, 184, 189, 191, 215
United Papermakers and Paperwork-
    ers, 219
United Paperworkers of America
    (CIO), 217, 218, 219
United Paperworkers International
    Union, 221
United States Conciliation Service,
    95
United Steelworkers of America, 6

Vernon, Calif., 163
Vigo County (Ind.) Central Labor
    Union, 88
Virginia, 193, 216

Wages, 20, 28, 29, 36–38, 49, 60–62, 69–70, 86, 109, 113, 139, 150, 151, 197–202, 214

Wagner Act, 94, 96; *see also* National Labor Relations Act, 1935–

Washington (state of), 135, 139

Wausau, Wis., 21, 80, 97, 134

Weinberg, A.N. (business agent, Local 299), 158, 159, 165, 206

Wenth, F.W. (local unionist), 79

Wentz, Keith (Pulp, Sulphite Workers' organizer), 119

West Virginia Pulp and Paper Company, 22, 89, 188, 216–17

Weyerhaeuser Timber Company, Pulp Division, 77, 91, 135

Wheelock, Ervin, 134

Williams, Willie B. (local unionist), 113

Wilson, W.H., 73

Wisconsin: Pulp, Sulphite Workers' activities in, 11, 68, 72–73, 93, 97–98, 119, 127, 128, 133, 134

Wisconsin Rapids, Wis., 134

Wisconsin River Valley, 72, 97–98

Wisconsin State Federation of Labor, 183

Wolf, Robert B., 142; support for Pulp, Sulphite Workers, 77, 135–37, 141

Women workers: dues of, 110, 208–11; in Pulp, Sulphite Workers, 106–10, 194, 208–9, 214; in paper industry, 29, 34, 36, 39, 107, 160–61, 208–11; wage differentials, 37, 86, 200–1

Women's Trade Union League of Chicago, 109, 162

World War I: impact on paper industry, 5, 16, 34

World War II: impact on paper industry unions, 5, 194, 199, 205, 213, 216–17

Wray, Harriet (Pulp, Sulphite Workers' organizer), 157, 165

Wray, Morris (Pulp, Sulphite Workers' organizer), 124, 125, 157, 160, 165, 183–84

York, Pa., 92

Zellerbach, J.D., 138, 139

Twentieth-Century America Series

DEWEY W. GRANTHAM, GENERAL EDITOR

Each volume in this series focuses on some aspects of the politics of social change in recent American history, utilizing new approaches to clarify the response of Americans to the dislocating forces of our own day—economic, technological, racial, demographic, and administrative.

The Reaffirmation of Republicanism: Eisenhower and the Eighty-third Congress by Gary W. Reichard

The Crisis of Conservative Virginia: The Byrd Organization and the Politics of Massive Resistance by James W. Ely, Jr.

Black Tennesseans, 1900–1930 by Lester C. Lamon

Political Power in Birmingham, 1871–1921 by Carl V. Harris

The Challenge to Urban Liberalism: Federal-City Relations during World War II by Philip J. Funigiello

Testing the Roosevelt Coalition: Connecticut Society and Politics in the Era of World War II by John W. Jeffries

Black Americans in the Roosevelt Era: Liberalism and Race by John B. Kirby

American Political History as Social Analysis: Essays by Samuel P. Hays

Suburb: Neighborhood and Community in Forest Park, Ohio, 1935–1976 by Zane L. Miller

Miners, Millhands, and Mountaineers: The Industrialization of the Appalachian South, 1880–1930 by Ronald D. Eller

Jonathan Daniels and Race Relations: The Evolution of a Southern Liberal by Charles W. Eagles

Southern Progressivism: The Reconciliation of Progress and Tradition by Dewey W. Grantham

THE UNIVERSITY OF TENNESSEE PRESS : KNOXVILLE